Österreichische Akademie der Wissenschaften

Mathematisch-naturwissenschaftliche Klasse

Sitzungsberichte

Abteilung I

Biologische Wissenschaften und Erdwissenschaften

212. Band
Jahrgang 2006

Wien 2007

Verlag der Österreichischen Akademie der Wissenschaften

Inhalt

Sitzungsberichte Abt. I

Sitzungsber. Abt. I (2006) 212: 3–18

Sitzungsberichte

Mathematisch-naturwissenschaftliche Klasse Abt. I
Biologische Wissenschaften und Erdwissenschaften

Altitudinal Variation in Flowering Time of Lilac (*Syringa vulgaris* L.) in the Alps in Relation to Temperatures

By

Walter Larcher

(Vorgelegt in der Sitzung der math.-nat. Klasse am 14. Dezember 2006
durch das w. M. Walter Larcher)

Abstract

The flowering time of lilac is a frequently used bioindicator for temperature driven developmental processes in spring time. In the present study the influence of altitude on the flowering dates of lilac was investigated for 24 sites in the Northern and Central Alps, and 10 sites in the Southern Alps. The onset of flowering dates in these areas of the Alps are closely related to the mean air temperatures in April. A temperature decrease by $1°C$ postpones the onset of flowering by 6.6 days. This results in a gradient of lilac flowering times of 4.2–4.6 days per 100 m in the Central Alps between 600 and 1500 m a.s.l. and a gradient of 3.7 days per 100 m a.s.l. in the Southern Alps between 90 and 1100 m a.s.l.

Over the 20th century a deviation from the mean onset of flowering dates of about ±12 days was seen due to climatic fluctuations in the Northern and Central Alps. On the Southern rim of the Alps the thermal conditions are most favourable for the development of lilac and therefore, there are no remarkable time shifts to be seen.

1. Introduction

Phenology, the science of the developmental changes in the course of a year, is becoming more and more important; especially in times when the changeability of the climate is strongly brought to people's attention. Developmental processes that respond to seasonal changes are the emergence of bud break, unfolding of leaves, onset of flowering and fruit ripening as well as leaf senescence. The evaluation of phenological dates provides valuable information about the average times of onset and the duration of different phenophases of characteristic species in a given area (ROSENKRANZ [29], SCHNELLE [30], LAUSCHER [17]).

Temperature and day length are the most important factors influencing the life cycle of plants. For trees and shrubs in the temperate Northern Hemisphere, spring temperatures play a decisive role (BORCHERT et al. [1]). Thus, spring phenology seems to be a promising indicator for the impact of climate warming (MENZEL and FABIAN [22], CHMIELEWSKI and RÖTZER [2]). However, the relationships between climatic factors and defined stages of development of certain indicator plants are still difficult to prove as too many factors overlap. Data have shown that correlations can only be seen for restricted areas with comparable climatic conditions (DEFILA and CLOT [6]).

The flowering time of lilac serves as an excellent bioindicator for temperature-driven developmental processes. The lilac bloom marks peak the spring time and can be observed effortlessly. IHNE [10] in his early phenological European maps used this widespread ornamental shrub as a signal for the onset of spring. SCHWARTZ et al. [32] used species of the genus *Syringa* as indicators of spring in a model for the Northern Hemisphere. In long-term surveys LAUSCHER and SCHNELLE [19] found a correlation between the onset of lilac flowering and temperature. Lilac bloom data were also used by ROLLER [28] and KOCH [11, 12] as an indicator for the altitudinal gradient of temperatures.

The following case study investigated the relationships between the altitudinal gradient of air temperatures and the impact of altitude on onset of flowering of lilac at latitudes between 48° and −46° N.

2. Materials and Methods

2.1. Acquisition of Data

Phenological Data. The phenological data presented were mainly compiled by literature research (*see* Table 1 and Fig. 4) and were completed with my own observations in Innsbruck (2003–2006) and in Arco, north of Lake Garda (1995–2004). The survey in Arco was carried out in an Arboretum in cooperation with the Natural Science Museum of Trento as part of the "Italian Phenological Garden Network" (TISI et al. [33]).

Phenological data series from various sites in the *Northern and Central Alps* in the Tyrol (Kufstein, Scharnitz, Telfs and Matrei in Osttirol) were taken from long-term observations (1951–1999) of the Central Institute for Meteorology and Geodynamics in Vienna (KOCH [13]). Furthermore, phenological data from Kufstein, Innsbruck, Landeck, Lienz, Matrei in Osttirol and St. Jakob in Defereggen from between 1946 and 1960 were also taken from the Central Institute for Meteorology and Geodynamics in Vienna (ROLLER [27]). The period between 1961 and 1973 is covered by annotations of 19 locations provided by a short term programme of a Phenological Observation Service in the Tyrol.

Earlier data series from sites in the *Southern Alps* between 1922 and 1961 were obtained from PFAFF [25], DALLA FIOR [4, 5] and MINIO [23]. The sites investigated were Bozen, Oberbozen and Eppan in Southern Tyrol, as well as Trento, Borgo (Val Sugana), Cles (Val di Sole), Tione (Giudicarie), Pieve di Ledro and Predazzo (Val di Fiemme) in Trentino. Some data covering the period before 1900 from the Tyrol and Trentino were taken from lists of DALLA TORRE in FICKER [8]. I visited all the sites.

Climate Data. Air temperature data were taken from recordings carried out by various weather services. For sites in the Northern and Central Alps these data were obtained from the Central Institute for Meteorology and Geodynamics in Vienna (www.ZAMG.ac.at) and from the Hydrographical Service in Austria. Older time series were taken from FICKER [8] and REITER [26]. Temperature recordings for the Southern Alps were researched on the Internet; for Southern Tyrol at www.provinz.bz.it/wetter (Hydrographical Service, Province Bozen) and for Trentino at www.ismaa.it (Istituto Agrario San Michele all'Adige). The climate data for Arco were obtained from a local weather station (LARCHER [16] and own data until 2004).

2.2. Processing of Data

The onset dates of characteristic phenological phases were established from the observations and were recorded as day of the year. These data are to be seen as means, which can vary by at least ±2–3 days as the observations were not carried out on a daily basis.

Onset of Flowering. The onset of lilac flowering is officially defined by the German Weather Service as the time when "*the lowest blossoms on numerous panicles of the lilac shrub start opening*" (DEUTSCHER WETTERDIENST [7, p. 13]). The phenological Observation Service for the Tyrol also gives a day of the year for the onset of lilac flowering. The phenological observers in the Southern Alps also use the same methodology (DALLA FIOR [3]: *"inizio dell'antesi"*; MARCELLO [20]: *"boccioli rigonfi e fiori aperti"*).

Phenological Calculations. Means, standard deviations, medians and percentiles were calculated from the long-term data taken from publications by KOCH [13] to give. In the case of all the other data only means and some standard deviations were calculated. The relationship between flowering dates and temperatures was identified by correlation analysis (Statistical package SPSS Inc., Chicago, USA).

To calculate the relationships between phenological information and temperature, the same periods had to be used. During the phenological observation period considerable temperature fluctuations took place (Fig. 1). A series of air temperature data from Innsbruck (115 years) shows colder periods before 1900, between 1935 and 1941 and between 1954 and 1987; warmer periods were registered between 1911 and 1934 as well as in the years between 1943 and 1951, with a strong warming after 1988. The warmest year of the 20th century was 1994. It is remarkable that the spring temperature means do not

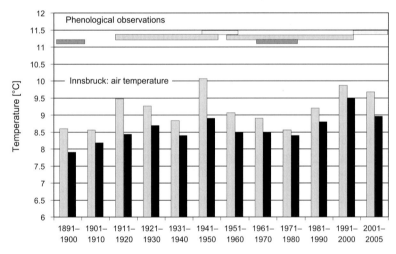

Fig. 1. Phenological observation periods and air temperatures in Innsbruck. *Bars*: phenological data from 1889–1900 (DALLA TORRE in FICKER [8]), 1910–1950 (data from Southern Tyrol and Trentino, *see text*), 1946–1960 (ROLLER [27]), 1951–1999 (KOCH [13]), 1961–1973 (Phenological Observation Service for Northern and Eastern Tyrol). Long-term data series of air temperature in Innsbruck: *black columns*: annual means, *gray columns*: spring means (March, April and May). The annual mean for the 115-year study period was 8.6°C and the spring mean 9.2°C

always match the annual means. This was especially striking in the decades between 1911–1930 and 1943–1951 with the warmest spring temperatures in 1946 and 1947. A similar trend could be seen for the Southern Alps (Bozen: warm periods 1926–1930 and 1941–1960).

3. Results

3.1. The Sequence of Phenophases of Lilac

During the course of a year plants undergo a series of developmental phases: Towards the end of the summer next year's buds are formed in the axils of the leaves; flower buds of lilac start differentiating as early as July (Fig. 2). Before leaves are shed the buds are rendered dormant by hormones, which prevent them from opening before the onset of winter. The termination of dormancy depends on the fulfilment of certain chilling requirements. In most cases, dormancy can only end after exposure to lower temperatures of about 6–8°C for a number of weeks. For lilac the cold requirement is usually fulfilled by the end of November (SCHÜEPP [31]). After that, lilac twigs flower after being exposed to warm water and a room temperature of 15–18°C (MOLISCH [24, p. 202]).

In the temperate climate zone the timing of bud break and flowering mostly depends on passing a certain temperature threshold. When the threshold is passed, elongation growth of the inflorescence axis is set off. Right after bud break the growth is rather slow. Only after sufficient temperature input is the programmed and quicker elongation growth achieved. So for elongation growth it is necessary

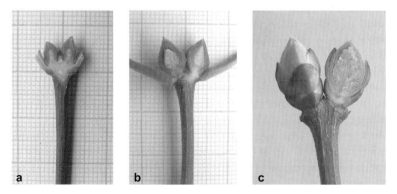

Fig. 2. Flower buds of lilac: (a) mid-July in the year before blooming; (b) mid-September; (c) before bud break in end of March of the current year

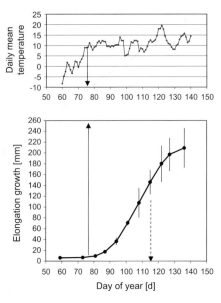

Fig. 3. Sprouting of flower buds and growth of inflorescences of *Syringa vulgaris*, and daily means of air temperature in Innsbruck between March 1 and May 20, 2005. The lilac hedge is located 2 km from the weather service's meteorological station. *Error bars*: standard deviation ($n = 30$). *Left arrows*: first day of bud break. *Right arrow*: onset of flowering. The data show a typically sigmoid course as the growth starts off slowly, gets faster in the linear elongation phase and ceases after having reached the maximum. During the phase of elongation growth daily means of about 5–10°C were predominant. At the onset of flowering daily mean temperatures in Innsbruck were higher than 10°C

that a certain temperature threshold is passed to start the flowering period and also that the plant is exposed to suitable temperatures over a certain period of time. The moment the first flowers of the panicle start flowering, growth of the inflorescence slows down. Fig. 3 shows that there was an interval of about 40 days between the first day of sprouting and the onset of flowering. The linear elongation phase from the start to the peak of flowering lasted about 50 days.

In Innsbruck lilac needs about 30–35 days from anthesis through peak flowering to the end of the flowering period; fruit maturity and capsule dehiscence start about 120–130 days after anthesis. Under optimal temperature conditions, such as those found in Arco on the southern rim of the Southern Alps, anthesis lasts about 25 days and fruit dehiscence is reached after 100–110 days.

3.2. Onset of Lilac Flowering at Higher Altitudes
of Mountain Ranges

Increasing altitude reduces the temperature and thus postpones the onset of spring. Dates for the onset of flowering of lilac at different locations are listed in Table 1 and Fig. 4. These phenological data show a gradient of 4.2–4.6 days per 100 m for the Central Alps between 600 and 1500 m a.s.l., and a gradient of 3.7 days per 100 m for the Southern Alps between 90 and 1100 m a.s.l.

Fig. 5 shows the shift in onset of flowering dates with elevation in the different areas of the north-south-transect. In comparison to the Central Alps, the onset dates at all altitudes on the northern rim of the Alps are delayed and the onset dates in the Southern Alps are earlier. A shift towards a higher day number indicates a dominant influence

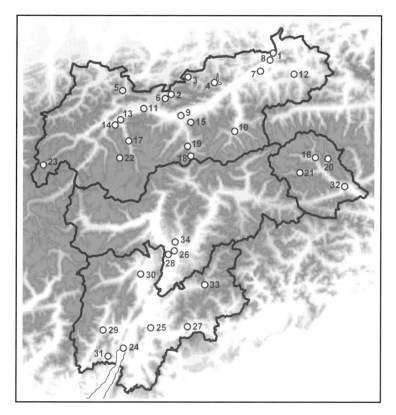

Fig. 4. Phenological stations for lilac flowering data. *Numbers see Table 1.* (Map showing topography: Dr. E. TASSER)

Table 1. Onset of flowering dates of *Syringa vulgaris* at different altitudes in locations within Tyrol and Trentino. Data taken from (a) Phenological Observation Service for Tyrol; (b) DALLA FIOR [4]; (c) DALLA FIOR [5]; (d) DALLA TORRE in FICKER [8]; (e) J. KERER, pers. comm.; (f) KOCH [13]; (g) KÖCK and TRENKWALDER [14]; (h) MINIO [23]; (i) PFAFF [25]; (k) ROLLER [27]; (l) own data

Location		Altitude [m]	Day of year [d]	Date	Phenological observation	Ref.
Northern Alps						
1	Kufstein	492	131	May 11	1951–1999	f
2	Scharnitz	964	155	June 04	1951–1999	j
3	Hinterriß	930	159	June 08	1961–1973	a
4	Pertisau	933	152	June 01	1961–1973	a
5	Ehrwald	1000	149	May 29	1961–1973	a
6	Leutasch	1126	168	June 17	1961–1973	a
Central Alps and Inn Valley						
7	Radfeld	510	128	May 08	1961–1973	a
8	Kirchbichl	550	136	May 16	1961–1973	a
9	Innsbruck	580	124	May 04	1889–1900	d
9a	Innsbruck	579	116	April 26	2003–2006	l
10	Zell am Ziller	585	131	May 11	1961–1973	a
11	Telfs	634	129	May 09	1951–1999	f
12	Kitzbühel	760	140	May 20	1961–1973	a
13	Zams	772	134	May 14	1961–1973	a
14	Landeck	818	126	May 06	1946–1960	k
14a	Landeck	825	138	May 18	1961–1973	a
15	Rinn	900	144	May 24	1972–1988	g
16	Matrei in Osttirol	975	146	May 26	1951–1999	f
16a	Matrei in Osttirol	975	140	May 20	1946–1960	k
17	Umhausen	1036	142	May 22	1961–1973	a
18	Brenner-Gries	1157	160	June 09	1951–1999	a
19	Trins	1214	157	June 06	1961–1973	a
20	Kals am Großglockner	1347	162	June 11	1961–1966	a
20a	Kals am Großglockner	1347	153	June 02	1995–2006	e
21	St. Jakob in Defereggen	1389	156	June 05	1946–1960	k
22	St. Leonhard in Pitztal	1371	167	June 16	1961–1973	a
23	Galtür	1583	173	June 22	1961–1973	a
Southern Alps						
24	Arco	91	100	April 10	1995–2004	l
25	Trento	210	106	April 16	1922–1951	b
26	Bozen	254	107	April 17	1915–1943	b
27	Borgo Val Sugana	384	110	April 20	1922–1961	c
28	Eppan	400	114	April 24	1922–1941	b
29	Tione Giudicarie	565	123	May 03	1922–1961	c
30	Cles	658	130	May 10	1922–1951	h
31	Pieve di Ledro	659	128	May 08	1922–1961	c
32	Lienz	666	122	May 02	1946–1960	k
33	Pedrazzo	1104	134	May 14	1922–1951	b
34	Oberbozen	1200	143	May 23	1916–1918	i

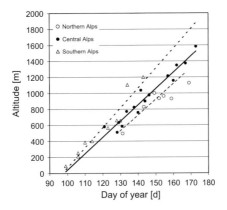

Fig. 5. Changes in flowering times (day of year) of *Syringa vulgaris* with increasing altitude along the north-south-transect. *Circles*: Northern Alps. *Dots*: Inn valley and valleys of the Central Alps. *Triangles*: Southern Alps

of temperature. Northern mountain ranges are cooler and rich in precipitation due to the Atlantic influence whereas inner alpine areas and mountain masses like the Central Alps are warmer. The locations in the Southern Alps are drier and warmer in spring (FLIRI [9], KUHN [15]). There is a close correlation between the flowering dates of all the sites in the north-south-transect and the mean temperature of the representative spring month April (Fig. 6). All in all a reduction in temperature of 1°C postpones the onset of flowering by 6.6 days.

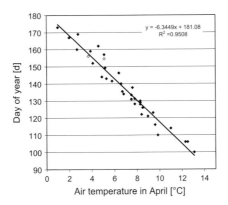

Fig. 6. Correlation between onset of flowering dates at all locations and mean temperatures in April. The relationship is highly significant ($r = -0.975$; $p < 0.001$; PEARSON)

3.3. Threshold Temperatures for Onset of Lilac Flowering

Bud break and onset of flowering are only possible if temperatures regularly and consistently pass a specific threshold temperature. In 2005 in Innsbruck, daily means during linear elongation growth were $10 \pm 2°C$. During peak flowering they were 10–15°C with minimum daily means never below 5°C (*see* Fig. 3).

When looking for any correlation between phenological dates and temperature, the mean temperature of a 30-day period before the onset of the given phase (LAUSCHER and SCHNELLE [19]) is an easy option. The month before the onset of a certain phase best demonstrates the relationship between phenological dates and temperature.

To reveal relationships between the onset of flowering dates of lilac and site temperatures, daily means of the preceding 30 days must be known. In most cases daily temperatures were not available, therefore mean temperatures of the 30 days preceding flowering were calculated as follows: for onset of flowering at the beginning of a month (e.g. May), the monthly mean of the previous month (e.g. April) was taken; when flowering started in the middle of a month, the mean temperature of the *flowering month* (T_{FM}) and that of the *previous month* (T_{PM}) were summed and divided by 2; when flowering started 10 days into a month, the mean temperature (Tm) was calculated by the formula $Tm = (1 * T_{FM} + 2 * T_{PM})/3$, when flowering started 20 days into the month the formula $Tm = (1 * T_{PM} + 2 * T_{FM})/3$ was used.

In the Alps clear relationships could be shown between the onset of lilac flowering and the monthly means of the previous month (Table 2). LAUSCHER and SCHNELLE [19] used long-term phenological data sets to calculate a mean temperatures for the 30 days before the onset of flowering of 9.4°C for Paris and of 10.3°C for St. Petersburg. This leads to the conclusion that the threshold temperature is about 9–10°C.

Table 2. Average threshold temperatures and deviations (°C) of the month before the onset of lilac flowering in the Tyrolean Alps and Trentino Alps

Region in the Alps	Mean	Median	SD	Max	Min
Northern Alps	9.7	9.7	0.83	10.8	8.6
Central Alps and Inn valley	9.6	9.6	0.79	10.9	8.3
Southern Alps	9.7	9.8	0.69	10.6	8.4

4. Discussion

4.1. Temperature Requirements for the Flowering of Lilac

Successful development from flower bud initiation to unfolding of inflorescences depends on induction by environmental signals like temperature and photoperiod. For flowering process lilac requires the following temperature conditions: Towards the end of the previous year the buds have to be exposed to some weeks with temperatures below 6–8°C (*chilling requirement*). For bud break the temperature threshold of 9–10°C should be passed in March or April in the following year. Seasonal changes in insolation and changed photoperiod may interact with the temperatures.

Syringa vulgaris originally comes from Southeast Europe and therefore is favoured by temperatures in warmer and more temperate climates. On the Greek island Samos (Marathonkampos, 37°40' N, 300 m a.s.l.) some shrubs of lilac, which must have started flowering about a week before, were observed to be at the peak of flowering on April 18, 2005 (day of year about 100; CH. KÖRNER, pers. comm.). In this region the daily mean temperatures range from 8–12°C in winter (November, December, January) with a daily minimum between 3 and 8°C. During spring (March and April) means of 12–14°C were recorded. In Rome (41°53' N, 51 m a.s.l.; mean temperatures in March are 11.5°C and in April 14.4°C) lilac starts flowering around April 11 (day of year 101). In Arco at Lake Garda the average onset of flowering also takes place around April 10 (average day of year 100). Long-term data sets show that in Arco daily temperature minima below 5°C occur regularly between November and January. As early as March (monthly mean: 9.7°C) daily means are consistently above 10°C in Arco. These optimal growing conditions lead to the early onset of flowering of lilac.

Unfavourable conditions like cooler temperatures and longer snow cover at higher altitudes delay phenological dates. The onset of flowering is postponed due to later bud break and longer elongation growth. The collective data for Austria before the middle of the 20th century generally shows an altitudinal gradient of flowering dates from 3.1 to 4.7 days per 100 m (ROSENKRANZ [29]). For the period from 1961 to 1977 the means for the onset of flowering dates give an altitudinal gradient of 3.6 day/100 m between 100 and 1400 m a.s.l. (ROLLER [28]). During extremely warm years development happens much quicker in the mountains than in the valley resulting in a lower gradient of 2–2.5 day/100 m (KOCH [11, 12]).

Lilac cannot flower after either too mild a winter or insufficient warmth in spring. If temperatures remain below the threshold of 10°C, reproductive growth will be suppressed. Too short a growing season could hinder the differentiation of next year's buds. In the Alps lilac can be planted up to an altitude of 1500–1600 m a.s.l. where it flowers between the middle of June and the beginning of July (day of year 165–180) with daily means of 10–12°C. Towards the Pole (e.g. in Norway) lilac is planted up until 60° N (LAUSCHER [18]). These shrubs also start flowering at temperatures of 9.7–11.7°C between the beginning and middle of June (day of year 140–160). Where these temperatures are not reached successful planting cannot be expected.

In climate zones with warm winters the chilling requirements of *Syringa vulgaris* (of below 8°C) cannot be fulfilled and therefore primary flowering induction cannot take place. In a Tenerife garden (Los Rodeos, 28°28′ N; about 600 m a.s.l.) shrubs of lilac were planted but over a period of several years they have not flowered (M. S. JIMÉNES and D. MORALES, pers. comm.). At this site the coldest monthly mean is measured at 12.4°C and the minimum monthly mean at 9.2°C in February (from between 1944–1989, Instituto Nacional de Meteorologia, Centro Zonal de Santa Cruz). During winter permanent temperatures below 10°C hardly occur and thus lilacs only grow in their vegetative form. It is remarkable that among the 378 sites in North America and Eurasia mentioned in the publication of SCHWARTZ et al. [32] there are no phenological observations of *Syringa* south of 30° latitude.

4.2. Variation in Onset of Flowering Dates
Due to Climate Changes

Analyses of phenological data from the European International Phenological Gardens between the years 1959 and 1996 showed that a temperature rise by 1 degree centigrade results in leafing 6.3 days earlier (MENZEL [21]). Phenological phase displacements of −7 days due to a 1 degree centigrade rise in spring temperature were discovered for different trees and shrubs by CHMIELEWSKI and RÖTZER [2].

It would be informative to know the flowering dates of lilac in different regions in relation to climate change. The present survey however does not suffice for analysis, as there were too few sites and little continuous data. However, two locations north and south of the Alps, namely Kufstein and Matrei in Osttirol, were chosen

Table 3. Range of flowering dates of lilac from 1951 to 2000 in Kufstein and Matrei in Osttirol

Location	Years	Date	Day of Year	Periods
Kufstein	1951–1999	May 11	131	mean
Kufstein	1962–1973	May 24	144	cooler
Kufstein	1989–2000	May 01	121	warmer
Matrei in Osttirol	1951–1999	May 26	146	mean
Matrei in Osttirol	1962–1973	June 06	157	cooler
Matrei in Osttirol	1989–2000	May 14	134	warmer

for further analysis. From the middle of the last century onwards there were clear fluctuations between colder and warmer periods (*see* Fig. 1). When compared to the flowering dates of lilac from the long-term data series from 1951 to 1999 (KOCH [13]) the onset of flowering dates during a colder period (1962–1973) were late by 10 days in Kufstein and 11 days in Matrei; during warmer periods (1989–2000) the onset of flowering dates were 13 and 12 days earlier in the same locations (Table 3). A 3-week amplitude in onset of flowering dates can be detected in these regions within only one century.

On the southern rim of the Alps there are no remarkable phase displacements of the earliest flowering dates as the warmer temperatures for development and growth are always reached. During the early decades of the 20th century lilac started flowering in Bozen around the middle of April (day of year 105–107: PFAFF [25], DALLA FIOR [4]), which coincides precisely with observations in 2005. In Arco the average onset of flowering for lilac has not changed markedly over time although there have been variations due to weather anomalies especially frost and late winter drought (day of year median 99, percentile: $10\% = 92$; $90\% = 110$). However, temperatures over this long period have been quite consistent. The rise in temperature in Arco between 1960 and 2000 was 0.8°C, which means 0.2°C in 10 years (TISI et al. [33]). The average temperature rise in the Mediterranean region after 1960 was about 0.3°C per decade; north of the Alps the temperature rose by more than 0.7°C per decade (WALTHER et al. [34]). Based on this climate change data an earlier onset of flowering dates is not traceable for *Syringa vulgaris* growing on the rim of the Southern Alps. DEFILA and CLOT [6] did not find a tendency for earlier vegetation development in the region of the Rhone Valley and the southern side of the Alps. This is in contrast to the northern cantons of Switzerland.

Can lilac be used as a phenological indicator for climate change? Partly, but only in regions where the development of the plants is limited.

Acknowledgements

I thank the following institutions and persons for the supply of climate dates: Institute of Meteorology and Geophysics of the University Innsbruck (Prof. Dr. EKKEHARD DREISEITL), Central Institute for Meteorology and Geodynamics, Regional Center for Tirol and Vorarlberg (Dr. KARL GABL), Hydrographic Office of the Tyrolean Government, Innsbruck (Ing. MARTIN NEUNER). Important phenological data come from the former Phenological Service for Tyrol and from a project in collaboration with the Natural History Museum in Trento. I am very grateful to the numerous observers for collecting phenological data in the field.

References

[1] BORCHERT, R., ROBERTSON, K., SCHWARTZ, M. D., WILLIAMS-LINERA, G. (2005) Phenology of temperate trees in tropical climates. Int. J. Biometeorol. **50**: 57–65

[2] CHMIELEWSKI, F. M., RÖTZER, TH. (2001) Response of tree phenology to climatic change across Europe. Agric. For. Meteorol. **108**: 1001–1112

[3] DALLA FIOR, G. (1933) Un decennio di osservazioni fitofenologiche a Trento con riferimenti a quelle di altre stazioni della Venezia Tridentina. Ann. Istituto tecnico "Leonardo da Vinci" di Trento **1933**: 7–19

[4] DALLA FIOR, G. (1951) Un terzo decennio di osservazioni fitofenologiche a Trento e risultati di analoghe osservazioni compiute in altre stazioni del Trentino-Alto Adige. Studi trentini di Scienze Naturali **A28**: 3–32

[5] DALLA FIOR, G. (1963) Un quarto decennio di osservazioni fitofenologiche a Trento ed un secondo a Tione e a Pieve di Ledro, nonché osservazioni quinquennali, rispettivamente quadriennali, di un secondo decennio a Borgo e a Condino. Studi trentini di Scienze Naturali **A40**: 176–191

[6] DEFILA, C., CLOT, B. (2001) Phytophenological trends in Switzerland. Int. J. Biometeorol. **45**: 203–207

[7] DEUTSCHER WETTERDIENST: BUTTLER, K. P., SCHMID, W. (eds.) (1991) Anleitung für die phänologischen Beobachter des Deutschen Wetterdienstes, 3. Aufl. Deutscher Wetterdienst, Offenbach am Main

[8] FICKER, H. VON (1909) Klimatographie von Österreich IV: Klimatographie von Tirol und Vorarlberg (mit Zoo- und Phytobiologischen Beiträgen von K. W. v. DALLA TORRE), pp. 1–162. K.K. Zentralanstalt f. Meteorologie u. Geodynamik, Wien

[9] FLIRI, F. (1975) Das Klima der Alpen im Raume von Tirol. Wagner, Innsbruck

[10] IHNE, E. (1905) Phänologische Karte des Frühlingseinzugs in Mitteleuropa. Petermanns Geogr. Mitt. **5**: 97–108

[11] KOCH, E. (1993) Phänologische Jahresübersicht für Österreich im Jahr 1993. Wetter und Leben **45**: 91–93

[12] KOCH, E. (1994) Phänologische Jahresübersicht für Österreich im Jahr 1994. Wetter und Leben **46**: 237–240

[13] KOCH, E. (2002) Phänologie Österreichs. In: HARLFINGER, O., KOCH, E., SCHEIFINGER, H. (Hrsg.) Klimahandbuch der österreichischen Bodenschätze.

Klimatographie, Teil 2: Strahlung, Weinbau, Phänologie, pp. 156–242. Wagner, Innsbruck

[14] KÖCK, L., TRENKWALDER, K. (1989) Witterungsverlauf, dargestellt in meteorologischen und phänologischen Tabellen mit Angaben von Mittel- und Extremwerten in Dekaden und 30jährigen Durchschnittswerten des Beobachtungszeitraumes 1951–1980. In: KÖCK, L., HOLAUS, K. (eds.) 50 Jahre Landesanstalt für Pflanzenzucht und Samenprüfung in Rinn, pp. 197–210. Private edition, Innsbruck

[15] KUHN, M. (1997) Meteorologische und klimatische Bedingungen für die Flora von Nordtirol, Osttirol und Vorarlberg. In: POLATSCHEK, A. (eds.) Flora von Nordtirol, Osttirol und Vorarlberg. 1. Band, pp. 26–42. Tiroler Landesmuseum Ferdinandeum, Innsbruck

[16] LARCHER, W. (1978) Klima und Pflanzenleben in Arco. T.E.M.I., Trento

[17] LAUSCHER, F. (1978) Neue Analyse ältester und neuerer phänologischer Reihen. Arch. Met. Geoph. Biokl. **B26**: 373–385

[18] LAUSCHER, F. (1991) Durchschnittstemperaturen beim Eintritt phänologischer Phasen in Norwegen (self-published)

[19] LAUSCHER, F., SCHNELLE, F. (1986) Beiträge zur Phänologie Europas. V. Lange phänologische Reihen Europas und ihre Beziehungen zur Temperatur. Ber. Dtsch. Wetterdienst No. 169: 1–24

[20] MARCELLO, A. (1957) Il tempo e la stagione in fenologia. Nuovo Giornale Botanico Italiano **66**: 929–1034

[21] MENZEL, A. (2000) Trends in phenological phases in Europe between 1951 and 1996. Int. J. Biometeorol. **44**: 76–81

[22] MENZEL, A., FABIAN, P. (1999) Growing season extended in Europe. Nature **397**: 659

[23] MINIO, M. (1937) Le osservazioni fitofenologiche della Rete Italiana nel 1935. Nuovo Giornale Botanico Italiano **44**: 552–567

[24] MOLISCH, H. (1930) Physiologie als Theorie der Gärtnerei, 6. Aufl. G. Fischer, Jena

[25] PFAFF, W. (1920) IV. Über den Einfluss der Höhenlage auf den Eintritt der Vegetationsphasen. Phaenologische Mitteilungen (Hessen) **26**: 1–8

[26] REITER, E. R. (1958) Klima von Innsbruck 1931–1955. Statistisches Amt, Innsbruck

[27] ROLLER, M. (1963) Durchschnittswerte phänologischer Phasen aus dem Zeitraum 1946 bis 1960 für 103 Orte Österreichs. Wetter und Leben **15**: 1–12

[28] ROLLER, M. (1978) Neue Normalwerte der Höhenabhängigkeit phänologischer Phasen in den Ostalpen. Jb. Zentralanstalt für Meteorologie und Geodynamik Wien, Anh. **6**: 57–69

[29] ROSENKRANZ, F. (1951) Grundzüge der Phänologie. Fromme, Wien

[30] SCHNELLE, F. (1955) Pflanzenphänologie. Akad. Verlagsgesellschaft, Leipzig

[31] SCHÜEPP, W. (1950) Phänometrisches Experiment über die „Winterruhe" einiger Pflanzen. Wetter und Leben **2**: 205–211

[32] SCHWARTZ, M. D., AHAS, R., AASA, A. (2006) Onset of spring starting earlier across the Northern Hemisphere. Global Change Biology **12**: 343–361

[33] TISI, F., BRESCIANI, I., LARCHER, W. (2005) Monitoraggio fenologico in un parco di acclimatazione nell'Alto Garda. Informatore Botanico Italiano **37**(1B): 686–687

[34] WALTHER, G.-R., POST, E., CONVEY, P., MENZEL, A., PARMESAN, C., BEEBEE, T. J. C., FROMENTIN, J.-M., HOEGH-GULDBERG, O., BAIRLEIN, F. (2002) Ecological responses to recent climate change. Nature **416**: 389–394

Internet:
Istituto Agrario di San Michele all'Adige: www.ismaa.it
Hydrographisches Amt, Provinz Bozen: www.provinz.bz.it/wetter
Zentralanstalt für Meteorologie und Geodynamik: www.zamg.ac.at

Author's address: em. Prof. Dr. Dr. h.c. Walter Larcher, Institut für Botanik, Sternwartestrasse 15, 6020 Innsbruck, Austria. E-Mail: walter.larcher@uibk.ac.at.

Österreichische Akademie der Wissenschaften
Mathematisch-naturwissenschaftliche Klasse

Sitzungsberichte

Abteilung II

Mathematische, Physikalische
und Technische Wissenschaften

215. Band
Jahrgang 2006

Wien 2007

Verlag der Österreichischen Akademie der Wissenschaften

Inhalt

Sitzungsberichte Abt. II

Anzeiger Abt. II

(nach Seite 176 des Teiles Sitzungsberichte Abt. II)

Sitzungsber. Abt. II (2006) 215: 3–11

Sitzungsberichte
Mathematisch-naturwissenschaftliche Klasse Abt. II
Mathematische, Physikalische und Technische Wissenschaften

© Österreichische Akademie der Wissenschaften 2007
Printed in Austria

A Note on the Eigenvalues for Periodic Three-Dimensional Jacobi-Perron Algorithms

By

Fritz Schweiger

(Vorgelegt in der Sitzung der math.-nat. Klasse am 19. Januar 2006
durch das k. M. Fritz Schweiger)

Abstract

In their profound study on the connections between Lyapunov theory and approximation properties of Jacobi-Perron algorithm BROISE-ALAMICHEL and GUIVARC'H 2001 proved a generalization of an inequality due to PALEY and URSELL [2]. In this note this inequality is slightly refined for dimension $n = 3$. This shows that for the eigenvalues $\sigma_0 > |\sigma_1| \geq |\sigma_2| \geq |\sigma_3|$ of a periodic expansion the inequality $|\sigma_1 \sigma_2| < 1$ is true. Furthermore it allows a more direct proof for the inequality $\lambda_1 + \lambda_2 < 0$ where $\lambda_0 > \lambda_1 > \lambda_2 > \lambda_3$ are the Lyapunov exponents of the algorithm.

Mathematics Subject Classification (2000): 11K55, 11J13, 11J70.
Key words: Multidimensional continued fractions, periodic expansions, Lyapunov exponents.

1. Introduction

Let

$$T(x_1, x_2, x_3) = \left(\frac{x_2}{x_1} - a, \frac{x_3}{x_1} - b, \frac{1}{x_1} - c \right), \qquad a(x) = \left[\frac{x_2}{x_1} \right],$$

$$b(x) = \left[\frac{x_3}{x_1} \right], \qquad c(x) = \left[\frac{1}{x_1} \right]$$

denote the three-dimensional map related to Jacobi-Perron algorithm (PERRON [3], SCHWEIGER [4, 6]). Define

$$(a_s, b_s, c_s) = (a(T^{s-1}x), b(T^{s-1}x), c(T^{s-1}x)), \qquad 1 \leq s.$$

Note that the digits satisfy the so-called *Perron conditions*

$$0 \leq a_s \leq c_s, \qquad 0 \leq b_s \leq c_s, \qquad 1 \leq c_s.$$

Furthermore: If $a_s = c_s$ then $1 \leq b_{s+1}$. If $b_s = c_s$ then $a_{s+1} \leq b_{s+1}$. If $b_s = c_s$ and $a_{s+1} = b_{s+1}$ then $1 \leq a_{s+2}$.

We introduce the matrices

$$\beta^{(s)} := \begin{pmatrix} c_s & 0 & 0 & 1 \\ 1 & 0 & 0 & 0 \\ a_s & 1 & 0 & 0 \\ b_s & 0 & 1 & 0 \end{pmatrix}$$

and

$$\beta^{(1)} \cdots \beta^{(s)} := \begin{pmatrix} A_0^{(s+4)} & A_0^{(s+1)} & A_0^{(s+2)} & A_0^{(s+3)} \\ A_1^{(s+4)} & A_1^{(s+1)} & A_1^{(s+2)} & A_1^{(s+3)} \\ A_2^{(s+4)} & A_2^{(s+1)} & A_2^{(s+2)} & A_2^{(s+3)} \\ A_3^{(s+4)} & A_3^{(s+1)} & A_3^{(s+2)} & A_3^{(s+3)} \end{pmatrix}.$$

From this notation we read the recursion relation

$$A_\alpha^{(s+4)} = c_s A_\alpha^{(s+3)} + b_s A_\alpha^{(s+2)} + a_s A_\alpha^{(s+1)} + A_\alpha^{(s)}, \qquad \alpha \in \{0, 1, 2, 3\}.$$

The following expansion is worth to be stated as a separate lemma.

Lemma (BROISE-ALAMICHEL and GUIVARC'H [1]).

$$A_0^{(s+4)} = (c_s - 1)A_0^{(s+3)} + (c_{s-1} + b_s - 1)A_0^{(s+2)}$$
$$+ (c_{s-2} + b_{s-1} + a_s - 1)A_0^{(s+1)} + (c_{s-3} + b_{s-2} + a_{s-1} + 1)A_0^{(s)}$$
$$+ (b_{s-3} + a_{s-2} + 1)A_0^{(s-1)} + (a_{s-3} + 1)A_0^{(s-2)} + A_0^{(s-3)}.$$

We denote by

$$[s+i, s+j, s+k]_{\alpha,\beta} = \det \begin{pmatrix} A_0^{(s+i)} & A_0^{(s+j)} & A_0^{(s+k)} \\ A_\alpha^{(s+i)} & A_\alpha^{(s+j)} & A_\alpha^{(s+k)} \\ A_\beta^{(s+i)} & A_\beta^{(s+j)} & A_\beta^{(s+k)} \end{pmatrix}$$

the relevant determinants. Since the choice of $\alpha, \beta \in \{1, 2, 3\}$ is not important we drop these indices.

The following recursion relations are valid,

$$[s+4, s+1, s+2] = c_s[s+3, s+1, s+2] + [s, s+1, s+2],$$
$$[s+4, s+1, s+3] = b_s[s+2, s+1, s+3] + [s, s+1, s+3],$$
$$[s+4, s+2, s+3] = a_s[s+1, s+2, s+3] + [s, s+2, s+3].$$

These relations lead to the following useful expansion,

$$[s+4, s+2, s+3] = a_s[s+1, s+2, s+3] + b_{s-1}[s+1, s, s+2]$$
$$+ c_{s-2}[s+1, s-1, s] + [s-2, s-1, s].$$

We introduce the quantity

$$\rho_s := \frac{|[s, s-1, s-2]|}{A_0^{(s)}}, \qquad s \geq 4.$$

Theorem.

$$|[s+4, s+2, s+3]| \leq A_0^{(s+4)}\left(1 - \frac{A_0^{(s-2)}}{A_0^{(s+4)}}\right) \max_{s-2 \leq t \leq s+3} \rho_t.$$

Proof. The proof will be given by induction and by considering several cases. The relation

$$[s+4, s+2, s+3] = a_s[s+1, s+2, s+3] + b_{s-1}[s+1, s, s+2]$$
$$+ c_{s-2}[s+1, s-1, s] + [s-2, s-1, s]$$

immediately gives

$$|[s+4, s+2, s+3]| \leq (a_s A_0^{(s+3)} + b_{s-1} A_0^{(s+2)}$$
$$+ c_{s-2} A_0^{(s+1)} + A_0^{(s)}) \max_{s \leq t \leq s+3} \rho_t.$$

(1) If the following three conditions are satisfied, namely

$$a_s \leq c_s - 1, \qquad b_{s-1} \leq c_{s-1} + b_s - 1, \qquad c_{s-2} \leq c_{s-2} + b_{s-1} + a_s - 1,$$

then a comparison with

$$A_0^{(s+4)} = (c_s - 1)A_0^{(s+3)} + (c_{s-1} + b_s - 1)A_0^{(s+2)}$$
$$+ (c_{s-2} + b_{s-1} + a_s - 1)A_0^{(s+1)} + (c_{s-3} + b_{s-2} + a_{s-1} + 1)A_0^{(s)}$$
$$+ (b_{s-3} + a_{s-2} + 1)A_0^{(s-1)} + (a_{s-3} + 1)A_0^{(s-2)} + A_0^{(s-3)}$$

shows that

$$|[s+4, s+2, s+3]| \leq A_0^{(s+4)}\left(1 - \frac{A_0^{(s)}}{A_0^{(s+4)}}\right) \max_{s \leq t \leq s+3} \rho_t.$$

(2) Now assume that $c_s - 1 < a_s$. Then by Perron conditions we have $a_s = c_s$. Then we calculate

$$
\begin{aligned}
[s+4, s+2, s+3] &= c_s[s+1, s+2, s+3] + [s, s+2, s+3] \\
&= (c_s - 1)[s+1, s+2, s+3] \\
&\quad + [s+1, s+2, s+3] + [s, s+2, s+3] \\
&= (c_s - 1)[s+1, s+2, s+3] \\
&\quad + (a_{s-1} - b_{s-1})[s, s+1, s+2] \\
&\quad + (b_{s-2} - c_{s-2})[s, s-1, s+1] \\
&\quad + (1 + c_{s-3})[s, s-2, s-1] + [s-3, s-2, s-1].
\end{aligned}
$$

This gives the estimate

$$
\begin{aligned}
|[s+4, s+2, s+3]| \leq \max_{s-1 \leq t \leq s+3} \rho_t((c_s - 1)A_0^{(s+3)} + |a_{s-1} - b_{s-1}|A_0^{(s+2)} \\
+ |b_{s-2} - c_{s-2}|A_0^{(s+1)} + (1 + c_{s-3})A_0^{(s)} + A_0^{(s-1)}).
\end{aligned}
$$

We will show that the inequalities

$$
|a_{s-1} - b_{s-1}| \leq c_{s-1} + b_s - 1, \quad |b_{s-2} - c_{s-2}| \leq c_{s-2} + b_{s-1} + a_s - 1
$$

are satisfied. Then a comparison with the expansion in the lemma shows

$$
|[s+4, s+2, s+3]| \leq A_0^{(s+4)}\left(1 - \frac{A_0^{(s-2)}}{A_0^{(s+4)}}\right) \max_{s-2 \leq t \leq s+3} \rho_t.
$$

Suppose that $a_{s-1} - b_{s-1} > c_{s-1} + b_s - 1$. Then $c_{s-1} + 1 \geq a_{s-1} + 1 > c_{s-1} + b_{s-1} + b_s$. Then $b_{s-1} = b_s = 0$ and $c_{s-1} = a_{s-1}$. But the last condition implies $1 \leq b_s$, a contradiction.

If $b_{s-1} - a_{s-1} > c_{s-1} + b_s - 1$ then $c_{s-1} + 1 \geq b_{s-1} + 1 > c_{s-1} + a_{s-1} + b_s$. Hence $a_{s-1} = b_s = 0$ and $b_{s-1} = c_{s-1}$. But $b_{s-1} = c_{s-1}$ implies $a_s \leq b_s$, hence $a_s = 0$, a contradiction.

If $c_{s-2} - b_{s-2} > c_{s-2} + b_{s-1} + a_s - 1$ then $1 > b_{s-2} + b_{s-1} + a_s$. This leads to $b_{s-2} = b_{s-1} = a_s = 0$ which again contradicts $a_s = c_s$.

(3) Therefore from now on we assume $a_s \leq c_s - 1$. However, there are two conditions left which could be violated.

(3.1) We first assume $c_s \geq 2$.

(3.1.1) We consider the case

$$
b_{s-1} > b_s + c_{s-1} - 1.
$$

From $b_{s-1} \leq c_{s-1}$ we get $b_{s-1} = c_{s-1}$ and $b_s = 0$ and therefore also $a_s = 0$. Then we have the recursion

$$[s + 4, s + 2, s + 3] = b_{s-1}[s + 1, s, s + 2] + c_{s-2}[s + 1, s - 1, s]$$
$$+ [s - 2, s - 1, s].$$

Then we estimate

$$\|[s + 4, s + 2, s + 3]\| \leq \left(c_{s-1} A_0^{(s+2)} + c_{s-2} A_0^{(s+1)} + A_0^{(s)} \right) \left(\max_{s \leq t \leq s+2} \rho_t \right).$$

This expression must be compared with

$$A_0^{(s+4)} \geq (c_s - 1) A_0^{(s+3)} + (c_{s-1} - 1) A_0^{(s+2)}$$
$$+ (c_{s-2} + c_{s-1} - 1) A_0^{(s+1)} + (c_{s-3} + 1) A_0^{(s)}.$$

Since

$$c_{s-1} A_0^{(s+2)} \leq A_0^{(s+3)} + (c_{s-1} - 1) A_0^{(s+2)}$$

we obtain

$$\|[s + 4, s + 2, s + 3]\| \leq A_0^{(s+4)} \left(1 - \frac{A_0^{(s)}}{A_0^{(s+4)}} \right) \max_{s \leq t \leq s+3} \rho_t.$$

(3.1.2) Next suppose that

$$c_{s-2} > c_{s-2} + b_{s-1} + a_s - 1.$$

Then $a_s = b_{s-1} = 0$ and we find

$$[s + 4, s + 2, s + 3] = c_{s-2}[s + 1, s - 1, s] + [s - 2, s - 1, s].$$

This leads to a comparison of

$$c_{s-2} A_0^{(s+1)} + A_0^{(s)}$$

with

$$A_0^{(s+4)} \geq (c_s - 1) A_0^{(s+3)} + (c_{s-1} + b_s - 1) A_0^{(s+2)} + (c_{s-2} - 1) A_0^{(s+1)} + A_0^{(s)}.$$

Since

$$c_{s-2} A_0^{(s+1)} \leq A_0^{(s+3)} + (c_{s-2} - 1) A_0^{(s+1)}$$

we get the same estimate as before.

(3.2) The remaining case is $c_s = 1$.

(3.2.1)

$$b_{s-1} > b_s + c_{s-1} - 1$$

leads to $b_{s-1} = c_{s-1}$ and $b_s = 0$ and $a_s = 0$. We expand the relation

$$[s+4, s+2, s+3] = b_{s-1}[s+1, s, s+2] + c_{s-2}[s+1, s-1, s]$$
$$+ [s-2, s-1, s]$$

to obtain

$$[s+4, s+2, s+3] = (b_{s-1} - 1)[s+1, s, s+2]$$
$$+ (c_{s-2} - a_{s-2})[s+1, s-1, s]$$
$$- b_{s-3}[s-1, s-2, s] - c_{s-4}[s-1, s-3, s-2]$$
$$+ [s-2, s-1, s] - [s-4, s-3, s-2].$$

As before we compare this relation with the expansion

$$A_0^{(s+4)} = (c_s - 1)A_0^{(s+3)} + (c_{s-1} + b_s - 1)A_0^{(s+2)}$$
$$+ (c_{s-2} + c_{s-1} + a_s - 1)A_0^{(s+1)} + (c_{s-3} + b_{s-2} + a_{s-1} + 1)A_0^{(s)}$$
$$+ (b_{s-3} + a_{s-2} + 1)A_0^{(s-1)} + (a_{s-3} + 1)A_0^{(s-2)} + A_0^{(s-3)}.$$

Obviously, $c_{s-2} - a_{s-2} \le c_{s-2} + c_{s-1} + a_s - 1$ is true. The remaining critical estimate is

$$(b_{s-3} + 1)A_0^{(s)} + c_{s-4}A_0^{(s-1)} \le (c_{s-3} + b_{s-2} + a_{s-1} + 1)A_0^{(s)}$$
$$+ (b_{s-3} + a_{s-2} + 1)A_0^{(s-1)}$$

or equivalently

$$b_{s-3}A_0^{(s)} + c_{s-4}A_0^{(s-1)} \le (c_{s-3} + b_{s-2} + a_{s-1})A_0^{(s)} + (b_{s-3} + a_{s-2})A_0^{(s-1)}.$$

If $b_{s-3} = c_{s-3}$ then we obtain $a_{s-2} \le b_{s-2}$. If $b_{s-2} = 0$ then $a_{s-2} = 0$ and hence $a_{s-1} \ge 1$. The other case is $b_{s-1} \ge 1$. In both cases we obtain

$$c_{s-4}A_0^{(s-1)} \le A_0^{(s)} + A_0^{(s-1)}.$$

Hence

$$|[s+4, s+2, s+3]| \le A_0^{(s+4)}\left(1 - \frac{A_0^{(s-1)}}{A_0^{(s+4)}}\right) \max_{s-1 \le t \le s+3} \rho_t.$$

If $b_{s-3} < c_{s-3}$ then we get again

$$c_{s-4}A_0^{(s-1)} \le A_0^{(s)} + A_0^{(s-1)}$$

and the same result.

(3.2.2) If

$$c_{s-2} > c_{s-2} + b_{s-1} + a_s - 1,$$

then $b_{s-1} = a_s = 0$. Note that in this case clearly $b_{s-1} \leq b_s + c_{s-1} - 1$.
We look at

$$[s+4, s+2, s+3] = c_{s-2}[s+1, s-1, s] + [s-2, s-1, s]$$

and compare this with

$$A_0^{(s+4)} \geq (c_{s-2} - 1)A_0^{(s+1)} + (c_{s-3} + 1)A_0^{(s)} + (b_{s-3} + 1)A_0^{(s-1)}$$
$$+ (a_{s-3} + 1)A_0^{(s-2)} + A_0^{(s-3)}$$

which is equivalent to

$$A_0^{(s+4)} \geq c_{s-2}A_0^{(s+1)} + A_0^{(s)} + A_0^{(s-1)} + A_0^{(s-2)}.$$

This leads to the estimate

$$|[s+4, s+2, s+3]| \leq A_0^{(s+4)} \left(1 - \frac{A_0^{(s-1)}}{A_0^{(s+4)}} \right) \max_{s-2 \leq t \leq s+1} \rho_t.$$

For a periodic algorithm with eigenvalues $\sigma_0, \sigma_1, \sigma_2, \sigma_3$ ordered in a way such that $\sigma_0 > |\sigma_1| \geq |\sigma_2| \geq |\sigma_3|$ the following corollary follows.

Corollary 1. $|\sigma_1 \sigma_2| < 1$.

Proof. Note that for a periodic algorithm with period length p the relations

$$A_0^{(sp+4)} + A_1^{(sp+1)} + A_2^{(sp+2)} + A_3^{(sp+3)} \sim \sigma_0^s$$

and

$$[sp+4, sp+1, sp+2]_{0,1,2} + [sp+1, sp+2, sp+3]_{1,2,3}$$
$$+ [sp+4, sp+2, sp+3]_{0,2,3} + [sp+4, sp+1, sp+3]_{0,1,3}$$
$$\sim |\sigma_0 \sigma_1 \sigma_2|^s.$$

Here the quantities $[s+i, s+j, s+k]_{\alpha,\beta,\gamma}$ are defined as

$$[s+i, s+j, s+k]_{\alpha,\beta,\gamma} = \det \begin{pmatrix} A_\alpha^{(s+i)} & A_\alpha^{(s+j)} & A_\alpha^{(s+k)} \\ A_\beta^{(s+i)} & A_\beta^{(s+j)} & A_\beta^{(s+k)} \\ A_\gamma^{(s+i)} & A_\gamma^{(s+j)} & A_\gamma^{(s+k)} \end{pmatrix}.$$

Clearly the theorem of the paper extends to these numbers. Since the algorithm is periodic there is a constant $q<1$ such that

$$1 - \frac{A_0^{(s-2)}}{A_0^{(s+4)}} \leq q < 1.$$

If $\lambda_0, \lambda_1, \lambda_2, \lambda_3$ are the four Lyapunov exponents the following result can be proved by applying the ergodic theorem.

Corollary 2. $\lambda_0 + \lambda_3 > 0$.

Proof. The proof closely follows SCHWEIGER [5] (see also SCHWEIGER [4]). We introduce the quantity

$$\tau_s := \max_{0 \leq j \leq 5} \rho_{s+j}.$$

Then we find

$$\tau_{s+6} \leq \max_{0 \leq j \leq 5} \left(1 - \frac{A_0^{(s+j-2)}}{A_0^{(s+j+4)}}\right) \tau_s.$$

Using estimates like

$$1 - \frac{A_0^{(s-2)}}{A_0^{(s+4)}} \leq 1 - \frac{1}{4^6 c_s c_{s-1} c_{s-2} c_{s-3} c_{s-4} c_{s-5}},$$

we see that the product

$$\prod_{s=1}^{N} \left(1 - \max_{0 \leq j \leq 5} \left(1 - \frac{A_0^{(s+j-2)}}{A_0^{(s+j+4)}}\right)\right)^{1/N}$$

can be estimated almost everywhere by using the ergodic theorem. More precisely, there is a constant $\kappa < 1$ such that

$$\prod_{s=1}^{N} \left(1 - \max_{0 \leq j \leq 5} \left(1 - \frac{A_0^{(s+j-2)}}{A_0^{(s+j+4)}}\right)\right)^{1/N} \leq \kappa$$

holds almost everywhere. Since almost everywhere

$$\lim_{s \to \infty} \frac{\log A_0^{(s)}}{s} = \lambda_0,$$

the proof can be easily completed.

Acknowledgement

The author likes to thank B. SCHRATZBERGER who is engaged in the project P16964 of the Austrian Science Foundation (Multidimensional Continued Fractions) for helpful comments.

References

[1] BROISE-ALAMICHEL, A., GUIVARC'H, Y. (2001) Exposants caractéristiques de l'algorithme de Jacobi-Perron et de la transformation associée. Ann. Inst. Fourier (Grenoble) **51** (no. 3): 565–686

[2] PALEY, R. E. A. C., URSELL, H. D. (1930) Continued fractions in several dimensions. Proc. Cambridge Phil. Soc. **26**: 127–144

[3] PERRON, O. (1907) Grundlagen für eine Theorie des Jacobischen Kettenbruchalgorithmus. Math. Ann. **64**: 1–76

[4] SCHWEIGER, F. (2000) Multidimensional Continued Fractions. Oxford: Oxford University Press

[5] SCHWEIGER, F. (1996) The exponent of convergence for the 2-dimensional Jacobi-Perron algorithm. In: NOWAK, W. G., SCHOISSENGEIER, J. (eds.) Proceedings of the Conference on Analytic and Elementary Number Theory, pp. 207–213. Institut für Mathematik, Universität für Bodenkultur, Wien

[6] SCHWEIGER, F. (2002) Diophantine properties of multidimensional continued fractions. In: DUBICKAS, A., et al. (eds.) Analytic and Probabilistic Methods in Number Theory. Proc. 3rd International Conference in Honour of J. Kubilius, pp. 242–255. TEV Ltd., Vilnius

Author's address: Fritz Schweiger, Department of Mathematics, University of Salzburg, Hellbrunner Strasse 34, 5020 Salzburg, Austria. E-Mail: fritz.schweiger@ sbg.ac.at.

Sitzungsber. Abt. II (2006) 215: 13–35

Sitzungsberichte
Mathematisch-naturwissenschaftliche Klasse Abt. II
Mathematische, Physikalische und Technische Wissenschaften

© Österreichische Akademie der Wissenschaften 2007
Printed in Austria

Cubic Helices in Minkowski Space

By

Jiří Kosinka and Bert Jüttler[*]

(Vorgelegt in der Sitzung der math.-nat. Klasse am 23. März 2006
durch das k. M. Hellmuth Stachel)

Abstract

We discuss space-like and light-like polynomial cubic curves in Minkowski (or pseudo-Euclidean) space $\mathbb{R}^{2,1}$ with the property that the Minkowski-length of the first derivative vector (or hodograph) is the square of a polynomial. These curves, which are called Minkowski Pythagorean hodograph (MPH) curves, generalize a similar notion from the Euclidean space (see FAROUKI [3]). They can be used to represent the medial axis transform (MAT) of planar domains, where they lead to domains whose boundaries are rational curves. We show that any MPH cubic (including the case of light-like tangents) is a cubic helix in Minkowski space. Based on this result and on certain properties of tangent indicatrices of MPH curves, we classify the system of planar and spatial MPH cubics.

Key words: Helix, Minkowski space, Minkowski Pythagorean hodograph curve.

1. Introduction

Cubic curves with constant slope in Euclidean space have thoroughly been investigated by WUNDERLICH [13]. For any given slope α, there exists exactly one cubic curve in three-dimensional Euclidean space – which is called the cubic helix – for which the ratio of curvature to torsion equals α. Its normal form for $\alpha = \pi/4$ is given by

$$\mathbf{c}(t) = (3t^2, t - 3t^3, t + 3t^3)^\top. \tag{1}$$

---

[*] Corresponding author

Cubic helices for other slopes α are obtained by a uniform scaling of the z-coordinate. According to WAGNER and RAVANI [11], cubic helices are the only cubics which are equipped with a rational Frenet-Serret motion. More precisely, the unit tangent, normal and binormal of the curve can be described by rational functions.

Pythagorean hodograph (PH) curves in Euclidean space were introduced by FAROUKI and SAKKALIS [4]. While the only planar PH cubic is the so-called Tschirnhausen cubic, FAROUKI and SAKKALIS [5] proved later that spatial PH cubics are helices, i.e. curves of constant slope. A classification of PH cubics in Euclidean space can be obtained by combining these results: Any PH cubic can be constructed as a helix with any given slope "over" the Tschirnhausen cubic.

Later, this notion was generalized to Minkowski (pseudo-Euclidean) space. As observed by MOON [10] and by CHOI et al. [1], Minkowski Pythagorean hodograph (MPH) curves are very well suited for representing the so-called medial axis transforms (MAT) of planar domains.

Recall that the MAT of a planar domain is the closure of the set containing all points (x, y, r), where the circle with center (x, y) and radius r touches the boundary in at least two points and is fully contained within the domain. When the MAT is an MPH curve, the boundary curves of the associated planar domain admit rational parameterizations. Moreover, rational parameterizations of their offsets exist too, since the offsetting operations correspond to a translation in the direction of the time axis, which clearly preserves the MPH property.

These observations served to motivate constructions for MPH curves. Interpolation by MPH quartics was studied by KIM and AHN [6]. Recently, it was shown that any space-like MAT can approximately be converted into a G^1 cubic MPH spline curve (KOSINKA and JÜTTLER [7]).

This paper analyzes the geometric properties of MPH cubics. As the main result, it is shown that these curves are again helices and can be classified, similarly to the Euclidean case.

The remainder of this paper is organized as follows. Section 2 summarizes some basic notions and facts concerning three-dimensional Minkowski geometry, MPH curves, and the differential geometry of curves in Minkowski space. Section 3 recalls some properties of helices in Euclidean space and it discusses helices in Minkowski space. Section 5 presents a classification of planar MPH cubics. Based on these results and using the so-called tangent indicatrix of a space-like curve we give a complete classification of spatial MPH cubics. Finally, we conclude the paper.

2. Preliminaries

In this section we summarize some basic concepts and results concerning Minkowski space, MPH curves and differential geometry of curves in Minkowski space.

2.1. Minkowski Space

The three-dimensional Minkowski space $\mathbb{R}^{2,1}$ is a real linear space with an indefinite inner product given by the matrix $G = \mathrm{diag}(1, 1, -1)$. The inner product of two vectors $\mathbf{u} = (u_1, u_2, u_3)^\top$, $\mathbf{v} = (v_1, v_2, v_3)^\top$, $\mathbf{u}, \mathbf{v} \in \mathbb{R}^{2,1}$ is defined as

$$\langle \mathbf{u}, \mathbf{v} \rangle = \mathbf{u}^\top G \mathbf{v} = u_1 v_1 + u_2 v_2 - u_3 v_3. \qquad (2)$$

The three axes spanned by the vectors $\mathbf{e}_1 = (1, 0, 0)^\top$, $\mathbf{e}_2 = (0, 1, 0)^\top$ and $\mathbf{e}_3 = (0, 0, 1)^\top$ will be denoted as the x-, y- and r-axis, respectively.

Since the quadratic form defined by G is not positive definite as in the Euclidean case, the square norm of $\mathbf{u}$ defined by $\|\mathbf{u}\|^2 = \langle \mathbf{u}, \mathbf{u} \rangle$ may be positive, negative or zero. Motivated by the theory of relativity one distinguishes three so-called "causal characters" of vectors. A vector $\mathbf{u}$ is said to be space-like if $\|\mathbf{u}\|^2 > 0$, time-like if $\|\mathbf{u}\|^2 < 0$, and light-like (or isotropic) if $\|\mathbf{u}\|^2 = 0$.

Two vectors $\mathbf{u}, \mathbf{v} \in \mathbb{R}^{2,1}$ are said to be orthogonal if $\langle \mathbf{u}, \mathbf{v} \rangle = 0$. The cross-product in the Minkowski space can be defined analogously to the Euclidean case as

$$\mathbf{w} = \mathbf{u} \bowtie \mathbf{v} = (u_2 v_3 - u_3 v_2, u_3 v_1 - u_1 v_3, -u_1 v_2 + u_2 v_1)^\top. \qquad (3)$$

Clearly, $\langle \mathbf{u}, \mathbf{u} \bowtie \mathbf{v} \rangle = 0$ for all $\mathbf{u}, \mathbf{v} \in \mathbb{R}^{2,1}$.

A vector $\mathbf{u} \in \mathbb{R}^{2,1}$ is called a unit vector if $\|\mathbf{u}\|^2 = \pm 1$. The hyperboloid of one sheet given by $x^2 + y^2 - r^2 = 1$ spanned by the endpoints of all unit space-like vectors will be called the unit hyperboloid $\mathcal{H}$.

Let u, v and w be three vectors in $\mathbb{R}^{2,1}$. A scalar triple product of u, v and w is defined as

$$[u, v, w] = \langle u, v \bowtie w \rangle. \qquad (4)$$

The scalar triple product in Minkowski space is the same as in Euclidean space, since the sign change in Minkowski inner and cross product cancels out. Therefore $[u, v, w] = \det(u, v, w)$.

A plane in Minkowski space is called space-, time- or light-like if the restriction of the quadratic form defined by G on this plane is positive definite, indefinite nondegenerate or degenerate, respectively. The type of a plane ρ can be characterized by the Euclidean angle α included between ρ and the xy plane. For light-like planes, $\alpha = \pi/4$.

2.2. Lorentz Transforms

A linear transform $L\colon \mathbb{R}^{2,1} \to \mathbb{R}^{2,1}$ is called a Lorentz transform if it maintains the Minkowski inner product, i.e. $\langle \mathbf{u}, \mathbf{v} \rangle = \langle L\mathbf{u}, L\mathbf{v} \rangle$ for all $\mathbf{u}, \mathbf{v} \in \mathbb{R}^{2,1}$. The group of all Lorentz transforms $\mathcal{L} = O(2,1)$ is called the Lorentz group.

Let $K = (k_{i,j})_{i,j=1,2,3}$ be a Lorentz transform. Then the column vectors $\mathbf{k}_1$, $\mathbf{k}_2$ and $\mathbf{k}_3$ satisfy $\langle \mathbf{k}_i, \mathbf{k}_j \rangle = G_{i,j}$, $i,j \in \{1,2,3\}$, i.e. they form an orthonormal basis of $\mathbb{R}^{2,1}$.

From $\langle \mathbf{k}_3, \mathbf{k}_3 \rangle = G_{3,3} = -1$ one obtains $k_{33}^2 \geq 1$. A transform K is said to be orthochronous if $k_{33} \geq 1$. The determinant of any Lorentz transform K equals to ± 1, and special ones are characterized by $\det(K) = 1$.

The Lorentz group $\mathcal{L}$ consists of four components. The special orthochronous Lorentz transforms form a subgroup $SO_+(2,1)$ of $\mathcal{L}$. The other components are $T_1 \cdot SO_+(2,1)$, $T_2 \cdot SO_+(2,1)$ and $T_1 \cdot T_2 \cdot SO_+(2,1)$, where $T_1 = \mathrm{diag}(1,1,-1)$ and $T_2 = \mathrm{diag}(1,-1,1)$. Let

$$R(\alpha) = \begin{pmatrix} \cos \alpha & -\sin \alpha & 0 \\ \sin \alpha & \cos \alpha & 0 \\ 0 & 0 & 1 \end{pmatrix}$$

and

$$H(\beta) = \begin{pmatrix} 1 & 0 & 0 \\ 0 & \cosh \beta & \sinh \beta \\ 0 & \sinh \beta & \cosh \beta \end{pmatrix} \tag{5}$$

be a rotation of the spatial coordinates x, y, and a hyperbolic rotation with a hyperbolic angle β, respectively. Any special orthochronous Lorentz transform $L \in SO_+(2,1)$ can be represented as $L = R(\alpha_1)H(\beta)R(\alpha_2)$.

The restriction of the hyperbolic rotation to the time-like yr-plane (i.e. to Minkowski space $\mathbb{R}^{1,1}$) is given by

$$h(\beta) = \begin{pmatrix} \cosh \beta & \sinh \beta \\ \sinh \beta & \cosh \beta \end{pmatrix}. \tag{6}$$

2.3. MPH Curves

A curve segment $\mathbf{c}(t) \in \mathbb{R}^{2,1}$, $t \in [a, b]$ is called space-, time- or light-like if its tangent vector $\mathbf{c}'(t)$, $t \in [a, b]$ is space-, time- or light-like, respectively.

Recall that a polynomial curve in Euclidean space is said to be a *Pythagorean hodograph* (PH) curve (cf. FAROUKI [3]), if the norm of its first derivative vector (or "hodograph") is a (possibly piecewise) polynomial. Following MOON [10], a *Minkowski Pythagorean hodograph* (MPH) curve is defined similarly, but with respect to the norm induced by the Minkowski inner product. More precisely, a polynomial curve $\mathbf{c} \in \mathbb{R}^{2,1}$, $\mathbf{c} = (x, y, r)^{\top}$ is called an MPH curve if

$$x'^2 + y'^2 - r'^2 = \sigma^2 \tag{7}$$

for some polynomial σ.

Remark 1. As observed by MOON [10] and CHOI et al. [1], if the medial axis transform (MAT) of a planar domain is an MPH curve, then the coordinate functions of the corresponding boundary curves and their offsets are rational.

Remark 2. As an immediate consequence of the definition, the tangent vector $\mathbf{c}'(t)$ of an MPH curve cannot be time-like. Also, light-like tangent vectors $\mathbf{c}'(t)$ correspond to roots of the polynomial σ in (7).

2.4. Frenet Formulas in Minkowski Space

This section introduces several facts from the differential geometry of curves in Minkowski space, cf. WALRAVE [12]. We consider a curve segment $\mathbf{c}(t) \in \mathbb{R}^{2,1}$. In order to rule out straight line and inflections, we suppose that the first two derivative vectors $\mathbf{c}'(t)$ and $\mathbf{c}''(t)$ are linearly independent. More precisely, points with linearly dependent vectors $\mathbf{c}'(t)$ and $\mathbf{c}''(t)$ correspond to inflections in the sense of projective differential geometry, which will be excluded. We distinguish three different cases.

Case 1. Consider a space-like curve $\mathbf{c}(s) \in \mathbb{R}^{2,1}$, i.e. $\|\mathbf{c}'(s)\| > 0$. We may assume that the curve is parameterized by its arc length, i.e. $\|\mathbf{c}'(s)\| = 1$. Then we define a (space-like) unit tangent vector $\mathbf{T} = \mathbf{c}'(s)$ of $\mathbf{c}(s)$.

Subcase 1.1. If the vector $\mathbf{T}'$ is space-like or time-like on some parameter interval, the Frenet formulas take the form

$$\mathbf{T}' = \kappa\mathbf{N},$$
$$\mathbf{N}' = -\langle\mathbf{N},\mathbf{N}\rangle\kappa\mathbf{T} + \tau\mathbf{B},$$
$$\mathbf{B}' = \tau\mathbf{N}. \tag{8}$$

The unit vectors $\mathbf{N}$ and $\mathbf{B}$ are the unit normal and binormal vector, $\kappa > 0$ and τ are the Minkowski curvature and torsion of $\mathbf{c}(s)$, respectively. The three vectors $\mathbf{T}$, $\mathbf{N}$ and $\mathbf{B}$ form an orthonormal basis.

Subcase 1.2. The vector $\mathbf{T}'$ of a space-like curve may be light-like at an isolated point, or within an entire interval. The two cases will be called *Minkowski inflections* and *inflected segments*, respectively. The Frenet formulas of a space-like curve within an inflected segment take the form

$$\mathbf{T}' = \mathbf{N},$$
$$\mathbf{N}' = \tau\mathbf{N},$$
$$\mathbf{B}' = -\mathbf{T} - \tau\mathbf{B}, \tag{9}$$

where $\langle\mathbf{T},\mathbf{N}\rangle = \langle\mathbf{T},\mathbf{B}\rangle = 0$, $\langle\mathbf{N},\mathbf{N}\rangle = \langle\mathbf{B},\mathbf{B}\rangle = 0$ and $\langle\mathbf{N},\mathbf{B}\rangle = 1$. In this situation, the Minkowski curvature evaluates formally to $\kappa = 1$. This subcase covers curves lying in light-like planes.

Case 2. Consider a light-like curve $\mathbf{c}(s) \in \mathbb{R}^{2,1}$, i.e. $\langle\mathbf{c}'(s),\mathbf{c}'(s)\rangle = 0$. It follows that $\langle\mathbf{c}'(s),\mathbf{c}''(s)\rangle = 0$ and thus $\mathbf{c}''(s)$ lies in a light-like plane. Therefore $\mathbf{c}''(s)$ is space-like (light-like vector $\mathbf{c}''(s)$ leads to an inflection). We may assume that the curve is parameterized by its so-called pseudo arc length, i.e. $\|\mathbf{c}''(s)\| = 1$. Then we have

$$\mathbf{T}' = \mathbf{N},$$
$$\mathbf{N}' = \tau\mathbf{T} - \mathbf{B},$$
$$\mathbf{B}' = -\tau\mathbf{N}, \tag{10}$$

where $\langle\mathbf{N},\mathbf{T}\rangle = \langle\mathbf{N},\mathbf{B}\rangle = 0$, $\langle\mathbf{T},\mathbf{T}\rangle = \langle\mathbf{B},\mathbf{B}\rangle = 0$ and $\langle\mathbf{T},\mathbf{B}\rangle = 1$. Again, the Minkowski curvature evaluates to $\kappa = 1$.

Case 3. Let us consider a time-like curve $\mathbf{c}(s) \in \mathbb{R}^{2,1}$ parameterized by its arc length, i.e. $\|\mathbf{c}'(s)\| = -1$. Then we define a (time-like) unit tangent vector $\mathbf{T} = \mathbf{c}'(s)$. As $\langle\mathbf{T},\mathbf{T}'\rangle = 0$, the vector $\mathbf{T}'$ lies in a space-like plane. Therefore, $\mathbf{T}'$ is always space-like. The Frenet formulas take the form

$$\mathbf{T}' = \kappa \mathbf{N},$$
$$\mathbf{N}' = \kappa \mathbf{T} + \tau \mathbf{B},$$
$$\mathbf{B}' = -\tau \mathbf{N}, \tag{11}$$

where $\mathbf{N}$ and $\mathbf{B}$ are the unit normal and binormal vector, $\kappa > 0$ and τ are the Minkowski curvature and torsion of $\mathbf{c}(s)$, respectively. The three vectors $\mathbf{T}$, $\mathbf{N}$ and $\mathbf{B}$ form an orthonormal basis.

Remark 3. In the remainder of the paper, the notion of inflection also includes Minkowski inflections.

The next result characterizes inflections.

Proposition 4. *Let $\mathbf{c}(t) \in \mathbb{R}^{2,1}$ be a space-like curve. Then $\mathbf{c}(t)$ has an inflection corresponding to $t \in I$ if and only if $\|\mathbf{c}'(t) \bowtie \mathbf{c}''(t)\|^2 = 0$ for $t \in I$. This includes the case of an isolated inflection point, where $I = \{t_0\}$.*

Proof. For the sake of brevity, we will omit the dependence on the parameter t. Let $\mathbf{T}$ and κ be the unit tangent vector and the curvature of $\mathbf{c}$. As in the Euclidean case, the Frenet formulas imply

$$\|\mathbf{c}' \bowtie \mathbf{c}''\|^2 = \|\mathbf{c}'\|^6 \|\mathbf{T} \bowtie \mathbf{T}'\|^2. \tag{12}$$

Firstly, let $\mathbf{c}$ have an inflection. Then $\mathbf{T}'$ is a light-like vector. As $\langle \mathbf{T}, \mathbf{T} \rangle = 1$, by differentiating we obtain $\langle \mathbf{T}, \mathbf{T}' \rangle = 0$. One can easily check that $\langle \mathbf{T}, \mathbf{T}' \rangle = 0$ implies $\|\mathbf{T} \bowtie \mathbf{T}'\|^2 = 0$ (the geometric argument is that the vectors $\mathbf{T}$, $\mathbf{T}'$ define a light-like plane). Finally, (12) yields that $\|\mathbf{c}' \bowtie \mathbf{c}''\|^2 = 0$.

Secondly, let $\|\mathbf{c}' \bowtie \mathbf{c}''\|^2 = 0$. From (12) we get that $\|\mathbf{T} \bowtie \mathbf{T}'\|^2 = 0$ or $\kappa = 0$ has to hold. Again, since $\langle \mathbf{T}, \mathbf{T}' \rangle = 0$, we can conclude that the vector $\mathbf{T}'$ is light-like. □

The formulas

$$\kappa(t) = \frac{\sqrt{|\langle \mathbf{c}'(t) \bowtie \mathbf{c}''(t), \mathbf{c}'(t) \bowtie \mathbf{c}''(t) \rangle|}}{\|\mathbf{c}'(t)\|^3} \tag{13}$$

and

$$\tau(t) = \frac{[\mathbf{c}'(t), \mathbf{c}''(t), \mathbf{c}'''(t)]}{|\langle \mathbf{c}'(t) \bowtie \mathbf{c}''(t), \mathbf{c}'(t) \bowtie \mathbf{c}''(t) \rangle|} \tag{14}$$

for the curvature and torsion of a space-like curve $\mathbf{c}(t)$ without inflections can be derived from Frenet formulas. The proof is similar to the Euclidean case.

2.5. Curves of Zero Curvature or Torsion

Let us take a closer look at curves in Minkowski space with curvature or torsion identically equal to zero. One can verify that a curve $\mathbf{c}(t)$ in Euclidean or Minkowski space, whose curvature vanishes identically, is contained within a straight line.

Definition 5. A curve in $\mathbb{R}^3$ or $\mathbb{R}^{2,1}$ is called a spatial curve if and only if it does not lie in a plane.

In the Euclidean space $\mathbb{R}^3$, a curve, which is not a straight line, is planar (non-spatial) if and only if its torsion is identically equal to zero. Analogously, one may ask for curves in $\mathbb{R}^{2,1}$ with vanishing torsion.

The answer to this question is not the same as in Euclidean case. In fact, $\tau \equiv 0$ is neither a necessary nor a sufficient condition for a curve to be planar in Minkowski space.

On the one hand, it can be shown that curves lying in light-like planes consist only of Minkowski inflections, hence they are planar without having vanishing torsion. (The Minkowski curvature formally evaluates to 1, and the torsion plays the role of the curvature.) These curves correspond to Subcase 1.2 of the Frenet formulas.

On the other hand, curves with vanishing torsion are described by the following result.

Proposition 6 (WALRAVE [12]). *If a curve $\mathbf{c}(t) \in \mathbb{R}^{2,1}$ has vanishing torsion, then it is a planar curve or a curve similar to the so-called W-null-cubic*

$$\mathbf{w}(s) = \frac{1}{6\sqrt{2}}(6s - s^3, 3\sqrt{2}s^2, 6s + s^3)^{\top}. \tag{15}$$

Proof (Sketch, see WALRAVE [12] for details). Consider a curve $\mathbf{c}(s) \in \mathbb{R}^{2,1}$ such that $\tau \equiv 0$, $\kappa \neq 0$. When $\mathbf{c}(s)$ is space-like or time-like, one can easily verify that the third derivative $\mathbf{c}'''(s)$ of $\mathbf{c}(s)$ is a linear combination of $\mathbf{c}'(s)$ and $\mathbf{c}''(s)$, which implies that $\mathbf{c}(s)$ is a planar curve. When $\mathbf{c}(s)$ is light-like, $\tau \equiv 0$ yields $\mathbf{c}(s) = (1/6\sqrt{2})(6s - s^3, 3\sqrt{2}s^2, 6s + s^3)^{\top}$.

Therefore, the only spatial curve in $\mathbb{R}^{2,1}$ (up to Minkowski similarities) with torsion identically equal to zero is the light-like curve (15), which we will refer to as the W-null-cubic. Note that $\mathbf{w}(s)$ is parameterized by its pseudo arc length. □

Remark 7. (1) The W-null-cubic is also an MPH curve, since any polynomial light-like curve is an MPH curve.

(2) Throughout this paper, *similar* refers to the Minkowski geometry, i.e., it means equal up to Lorentz transforms, translations, and scaling.

3. Helices in Minkowski Space

We start with a brief summary of some basic results from Euclidean space.

A helix in $\mathbb{R}^3$ is a spatial curve for which the tangent makes a constant angle with a fixed line. Any such line is called the axis of the helix. Lancret's theorem states that a necessary and sufficient condition for a spatial curve to be a helix in $\mathbb{R}^3$ is that the ratio of its curvature to torsion is constant. The proof of this theorem uses Frenet formulas and can be found in many textbooks on classical differential geometry, e.g. KREYSZIG [8].

In the Euclidean version of the Lancret's theorem the restriction to spatial curves rules out curves with vanishing torsion. However, as shown in Section 2.5, this is generally not the case in Minkowski space.

Definition 8. A curve $\mathbf{c}(t) \in \mathbb{R}^{2,1}$ is called a helix if and only if there exists a constant vector $\mathbf{v} \neq (0,0,0)^\top$ such that $\langle \mathbf{T}(t), \mathbf{v} \rangle$ is constant, where $\mathbf{T}(t)$ is the unit tangent vector of $\mathbf{c}(t)$. Any line, which is parallel to the vector $\mathbf{v}$, is called an axis of the helix $\mathbf{c}(t)$.

Proposition 9 (Lancret's Theorem in $\mathbb{R}^{2,1}$). *A spatial curve $\mathbf{c}(t) \in \mathbb{R}^{2,1}$ is a helix if and only if $\tau = \alpha\kappa$, where α is a real constant and κ, τ are the Minkowski curvature and torsion of $\mathbf{c}(t)$.*

Proof. Recall that $\alpha = 0$ (i.e. $\tau \equiv 0$) corresponds to the W-null-cubic as shown in Proposition 6. From the Frenet formulas for light-like curves it follows that the binormal vector $\mathbf{B}$ of the W-null-cubic is a constant vector and $\langle \mathbf{T}, \mathbf{B} \rangle = 1$. This implies that the W-null-cubic is a helix in $\mathbb{R}^{2,1}$.

Now, let us suppose that $\alpha \neq 0$. As the proof is analogous for all five different cases of curves and corresponding Frenet formulas, we provide the proof of Lancret's theorem for two of the cases only.

Let $\mathbf{c}'(t)$ be space-like and $\mathbf{c}''(t)$ not light-like and let $\mathbf{c}(t)$ be a helix in $\mathbb{R}^{2,1}$. Then there exists a constant vector $\mathbf{v} \neq (0,0,0)^\top$ such that $\langle \mathbf{T}, \mathbf{v} \rangle = \beta$, $\beta \in \mathbb{R}$. By differentiating this equation with respect to t and using Frenet formulas we obtain $\langle \mathbf{N}, \mathbf{v} \rangle = 0$ and thus $\mathbf{v} = a\mathbf{T} + b\mathbf{B}$, where $a, b \in \mathbb{R}$. Again, by differentiating we get $\mathbf{N}(a\kappa + b\tau) = 0$, which gives $\tau = -(a/b)\kappa$ ($b = 0$ implies that $\mathbf{c}(t)$ is a straight line).

Conversely, let $\tau = \alpha\kappa$. Then we choose the vector $\mathbf{v} = \mathbf{T} - (1/\alpha)\mathbf{B}$. By differentiating this equation with respect to t one obtains that $\mathbf{v}' = (0,0,0)^\top$, i.e. $\mathbf{v}$ is a constant vector. Moreover, $\langle \mathbf{T}, \mathbf{v} \rangle = \langle \mathbf{T}, \mathbf{T} - (1/\alpha)\mathbf{B} \rangle = 1$, which proves that $\mathbf{c}(t)$ is a helix in $\mathbb{R}^{2,1}$. $\square$

Remark 10. For the remainder of the paper we restrict ourselves to space-like and light-like helices only. In order to avoid confusion, we will call these curves *SL-helices*.

Proposition 11. *Any polynomial SL-helix in Minkowski space is an MPH curve.*

Proof. Let $\mathbf{c}(t)$ be a polynomial light-like helix. Clearly, any polynomial light-like curve is an MPH curve.

Now, let $\mathbf{c}(t)$ be a polynomial space-like helix. Then there exists a constant vector $\mathbf{v}$ such that $\langle \mathbf{T}, \mathbf{v} \rangle = \alpha$, where α is a real constant. One can easily verify that $\alpha = 0$ leads to a contradiction with $\mathbf{c}(t)$ being a helix, since the torsion of $\mathbf{c}(t)$ would be identically equal to zero (cf. Proposition 6). Therefore $\alpha \neq 0$.

The unit tangent vector of $\mathbf{c}(t)$ can be obtained from

$$\mathbf{T} = \frac{\mathbf{c}'(t)}{\sqrt{\langle \mathbf{c}'(t), \mathbf{c}'(t) \rangle}}.$$

By substituting $\mathbf{T}$ in the first equation we obtain

$$\frac{\langle \mathbf{c}'(t), \mathbf{v} \rangle}{\alpha} = \sqrt{\langle \mathbf{c}'(t), \mathbf{c}'(t) \rangle}. \tag{16}$$

As the curve $\mathbf{c}(t)$ is a polynomial curve, the left-hand side of (16) is a polynomial. Consequently, the right-hand side of (16) is a polynomial as well and hence $\mathbf{c}(t)$ is an MPH curve. □

4. Spatial MPH Cubics

In this section we discuss the connection between polynomial helices in Minkowski space and MPH curves.

4.1. Space-Like MPH Cubics

Proposition 12. *The ratio of curvature to torsion of a spatial space-like MPH cubic is constant. Consequently, spatial space-like MPH cubics are helices in* $\mathbb{R}^{2,1}$.

Proof. We will prove the proposition by a direct computation. Let $\mathbf{c}(t) = (x(t), y(t), r(t))^{\top} \in \mathbb{R}^{2,1}$ be a spatial space-like MPH cubic. Then there exist four linear polynomials (cf. MOON [10])

$$u(t) = u_0(1-t) + u_1 t, \qquad v(t) = v_0(1-t) + v_1 t,$$
$$p(t) = p_0(1-t) + p_1 t, \qquad q(t) = q_0(1-t) + q_1 t, \tag{17}$$

such that

$$
\begin{aligned}
x'(t) &= u(t)^2 - v(t)^2 - p(t)^2 + q(t)^2, \\
y'(t) &= -2(u(t)v(t) + p(t)q(t)), \\
r'(t) &= 2(u(t)q(t) + v(t)p(t)), \\
\sigma(t) &= u(t)^2 + v(t)^2 - p(t)^2 - q(t)^2.
\end{aligned}
\tag{18}
$$

Since $\mathbf{c}(t)$ is a space-like curve, we may (without loss of generality) assume that $\mathbf{c}'(0) = (1,0,0)^\top$, which implies $u_0 = 0$, $p_0 = 0$ and $q_0^2 - v_0^2 = 1$. Expressing the curve $\mathbf{c}(t)$ in Bézier form and computing the following scalar triple product yields that $\mathbf{c}(s)$ is a planar curve if

$$
[\mathbf{c}'(0), \mathbf{c}'(1), \mathbf{c}(1) - \mathbf{c}(0)] = (u_1 - p_1)(u_1 + p_1)(v_1 q_0 - q_1 v_0) = 0.
\tag{19}
$$

According to Proposition 4, $\mathbf{c}(t)$ has an inflection point if

$$
(u_1 - p_1)(u_1 + p_1)\sigma(t) = 0.
\tag{20}
$$

One can observe from (19) and (20) that $\mathbf{c}(t)$ has an inflection if it has a light-like tangent (or it is a planar curve). Thus spatial space-like MPH cubics have no inflections.

Applying formulas (13) and (14) to the curve $\mathbf{c}(t)$ gives that the curvature and torsion of $\mathbf{c}(t)$ are given by

$$
\kappa(t) = \frac{2\sqrt{|(u_1 - p_1)(u_1 + p_1)|}}{\sigma^2(t)}, \qquad \tau(t) = \frac{2(v_1 q_0 - q_1 v_0)}{\sigma^2(t)}. \tag{21}
$$

Consequently, the ratio of $\kappa(t)$ to $\tau(t)$ does not depend on t. $\qquad\square$

4.2. Light-Like MPH Cubics

Proposition 13. *Any spatial light-like cubic is similar to the W-null-cubic.*

Proof. Consider a spatial light-like polynomial curve $\mathbf{c}(t)$ of degree 3. Let $t = t(s)$ be a reparameterization of $\mathbf{c}(t)$ such that $\mathbf{c}(t(s))$ is parameterized by the pseudo arc length and let $\mathbf{T}(s)$, $\mathbf{N}(s)$ and $\mathbf{B}(s)$ be the tangent, normal and binormal vector and $\tau(s)$ the torsion of $\mathbf{c}(t)$. We denote by $\mathbf{c}'$ and $\dot{\mathbf{c}}$ the first derivative of $\mathbf{c}$ with respect to t and s, respectively. For the sake of brevity we omit the dependence on s. Then we have

$$
\mathbf{T} = \mathbf{c}'(t)\frac{dt}{ds}.
\tag{22}
$$

Three consecutive differentiations of (22) with respect to s and simplifications using Frenet formulas yield

$$\mathbf{N} = \mathbf{c}''(t)\left(\frac{dt}{ds}\right)^2 + \mathbf{c}'(t)\frac{d^2t}{ds^2},$$

$$\tau\mathbf{T} - \mathbf{B} = \mathbf{c}'''(t)\left(\frac{dt}{ds}\right)^3 + 3\mathbf{c}''(t)\frac{dt}{ds}\frac{d^2t}{ds^2} + \mathbf{c}'(t)\frac{d^3t}{ds^3},$$

$$\dot{\tau}\mathbf{c}'(t)\frac{dt}{ds} = \dot{\tau}\mathbf{T} = \mathbf{c}''''(t)\left(\frac{dt}{ds}\right)^4 + a_3(s)\mathbf{c}'''(t) + a_2(s)\mathbf{c}''(t) + \mathbf{c}'(t)\frac{d^4t}{ds^4},$$

$$(23)$$

where

$$a_3(s) = \left(\frac{dt}{ds}\right)^2\frac{d^2t}{ds^2} = (\dot{t})^2\ddot{t}$$

and $a_2(s)$ is a function of s.

Consider the scalar triple product $[\mathbf{T}, \mathbf{N}, \tau\mathbf{T} - \mathbf{B}]$. Using (22) and (23) one obtains

$$[\mathbf{T}, \mathbf{N}, \tau\mathbf{T} - \mathbf{B}] = [\mathbf{T}, \mathbf{N}, -\mathbf{B}] = [\mathbf{c}'(t), \mathbf{c}''(t), \mathbf{c}'''(t)]\left(\frac{dt}{ds}\right)^6. \quad (24)$$

Since the vectors $\mathbf{T}(s)$, $\mathbf{N}(s)$ and $\mathbf{B}(s)$ are linearly independent, the vectors $\mathbf{c}'(t)$, $\mathbf{c}''(t)$ and $\mathbf{c}'''(t)$ are linearly independent as well. Consequently, by comparing coefficients and due to the fact that $\mathbf{c}''''(t) = (0, 0, 0)^\top$, the third equation of (23) implies

$$a_3(s) = 0, \qquad a_2(s) = 0, \qquad \dot{\tau}\frac{dt}{ds} = \frac{d^4t}{ds^4}. \quad (25)$$

From $a_3(s) = 0$ one may conclude that $t = \alpha s + \beta$ and therefore the last equation of (25) implies that τ is constant.

Finally, we express the binormal vector $\mathbf{B}$ using the second equation of (23):

$$\mathbf{B} = \tau\mathbf{T} - \alpha^3\mathbf{c}'''(t),$$

$$\dot{\mathbf{B}} = \dot{\tau}\mathbf{T} + \tau\dot{\mathbf{T}} - \alpha^4\mathbf{c}''''(t) = \tau\mathbf{N}. \quad (26)$$

On the other hand, from the Frenet formulas we have that $\dot{\mathbf{B}} = -\tau\mathbf{N}$. Therefore, the torsion τ is identically equal to zero. Proposition 6 concludes the proof. □

4.3. Summary

In this section we will summarize the previously obtained results (see the scheme in Fig. 1).

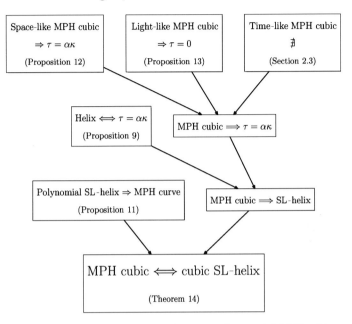

Fig. 1. Summary of obtained results concerning spatial MPH cubics

Theorem 14. *A spatial curve in Minkowski space is an MPH cubic if and only if it is a space-like or light-like cubic helix.*

Proof. From Propositions 12 and 13 it follows that any spatial MPH cubic satisfies the assumptions of the Lancret's theorem (Theorem 9) and therefore any such curve is a helix in Minkowski space. On the other hand, we have proved (cf. Proposition 11) that any polynomial SL-helix is an MPH curve. □

5. Classification of Planar MPH Cubics

In order to prepare the discussion of spatial helices, we present a classification of planar MPH cubics.

5.1. MPH Cubics in Space-Like and Light-Like Planes

A thorough discussion of planar MPH cubics in space-like planes was given in FAROUKI and SAKKALIS [4], since these curves are planar PH

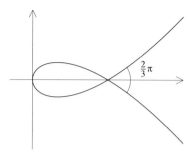

Fig. 2. Tschirnhausen cubic

cubics. It turns out that any planar PH cubic (which is not a straight line) is similar to the so-called Tschirnhausen cubic (see Fig. 2) given by $\mathbf{T}(t) = (3t^2, t - 3t^3)^\top$.

Let us consider a polynomial planar curve $\mathbf{c}(t) = (x(t), y(t))^\top$. The MPH condition in a light-like plane degenerates to $x'^2(t) = \sigma^2(t)$ and hence any polynomial curve in a light-like plane is an MPH curve. Therefore the only case remaining to consider is the case of a time-like plane.

5.2. MPH Cubics in Time-Like Planes

An MPH curve $\mathbf{c}(t) = (x(t), y(t))^\top$ lying in a time-like plane is nothing else but a curve in Minkowski plane $\mathbb{R}^{1,1}$ whose hodograph satisfies $x'^2(t) - y'^2(t) = \sigma^2(t)$, where $\sigma(t)$ is a polynomial in t.

Proposition 15. *Any MPH cubic in Minkowski plane $\mathbb{R}^{1,1}$ with exactly one point with a light-like tangent is similar to the curve $\mathbf{q}_1(t) = (t^3 + 3t, t^3 - 3t)^\top$ (depicted in Fig. 3).*

Any MPH cubic in Minkowski plane $\mathbb{R}^{1,1}$ with exactly two different points with light-like tangents is similar (in Minkowski sense) to the curve $\mathbf{q}_2(t) = (t^3 + 3t, 3t^2)^\top$ (see Fig. 4).

There are no MPH cubics in $\mathbb{R}^{1,1}$ except for the curves $\mathbf{q}_1(t)$, $\mathbf{q}_2(t)$ and straight lines.

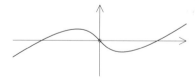

Fig. 3. MPH cubic in $\mathbb{R}^{1,1}$ with exactly one point with a light-like tangent (marked by the grey circle)

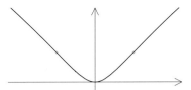

Fig. 4. MPH cubic in $\mathbb{R}^{1,1}$ with exactly two different points with light-like tangents (marked by the grey circles)

Proof. Let us suppose that $\mathbf{c}(t) = (x(t), y(t))^{\top}$ is an MPH cubic in $\mathbb{R}^{1,1}$. Then there exist two linear polynomials $u(t) = u_0(1-t) + u_1 t$, $v(t) = v_0(1-t) + v_1 t$ (cf. KUBOTA [9]) such that

$$x'(t) = u^2(t) + v^2(t),$$
$$y'(t) = 2u(t)v(t),$$
$$\sigma(t) = u^2(t) - v^2(t), \tag{27}$$

where σ is the parametric speed

$$\sigma = (at - u_0 - v_0)(bt - u_0 + v_0),$$

where

$$a = u_0 - u_1 + v_0 - v_1, \qquad b = u_0 - u_1 - v_0 + v_1. \tag{28}$$

Depending on the number of light-like tangents of $\mathbf{c}(t)$ we will distinguish the following four cases.

Case 1. The curve $\mathbf{c}(t)$ has infinitely many light-like tangents. This case occurs when $\sigma(t) \equiv 0$, which implies that $\mathbf{c}(t)$ is a part of a light-like straight line.

Case 2. The curve $\mathbf{c}(t)$ has no light-like tangents. This means that $\sigma(t)$ has no roots, i.e. $a = 0$ and $b = 0$, see Eq. (28). One can easily verify that $\mathbf{c}(t)$ is a part of a straight line.

Case 3. The curve $\mathbf{c}(t)$ has exactly one light-like tangent. The limit case, when the two roots of σ degenerate into one, gives again straight lines only. Let us (without loss of generality) suppose that $a \neq 0$ and $b = 0$. Then $\mathbf{c}(t)$ has a light-like tangent at $t_0 = (u_0 + v_0)/a$. We may assume that $t_0 = 0$ (otherwise we would reparameterize $\mathbf{c}(t)$) and therefore $u_0 + v_0 = 0$. A simple calculation reveals that

$$\mathbf{c}(t) = (\alpha t^3 + \beta t, \alpha t^3 - \beta t)^{\top}, \qquad \alpha = \tfrac{2}{3}(u_0 + v_1)^2, \qquad \beta = 2u_0^2. \tag{29}$$

By a reparameterization $t = t\sqrt{\beta/3\alpha}$ of $\mathbf{c}(t)$ given in (29) (the equality $\alpha = 0$ yields a straight line) and a scaling by factor $\sqrt{27\alpha/\beta^3}$ we obtain the curve $\mathbf{q}_1(t)$.

Case 4. The curve $\mathbf{c}(t)$ has exactly two different light-like tangents corresponding to $t_1 = (u_0 + v_0)/a$ and $t_2 = (v_0 - u_0)/b$, $a \neq 0, b \neq 0$ and $t_1 \neq t_2$. Let us consider the following transformation of the curve $\mathbf{c}(t) = (x(t), y(t))^\top$ consisting of a reparameterization $t = t + \delta$, a scaling by factor λ, a hyperbolic rotation with a hyperbolic angle φ and a translation given by the vector $(\varrho_1, \varrho_2)^\top$:

$$\mathbf{p}(t) = \lambda \begin{pmatrix} \cosh\varphi & \sinh\varphi \\ \sinh\varphi & \cosh\varphi \end{pmatrix} \mathbf{c}(t + \delta) + \begin{pmatrix} \varrho_1 \\ \varrho_2 \end{pmatrix}. \tag{30}$$

A straightforward but long computation gives that for the values

$$\varphi = \frac{1}{2}\ln\frac{b^2}{a^2}, \qquad \delta = \frac{u_0^2 - u_0 u_1 + v_0 v_1 - v_0^2}{ab}, \qquad \lambda = 3a^2\sqrt{\frac{b^2}{a^2}} \tag{31}$$

and ϱ_1, ϱ_2 such that $\mathbf{p}(0) = (0,0)^\top$, the curve becomes

$$\mathbf{p}(t) = (a^2 b^2 t^3 + 3c^2 t, 3abct^2)^\top, \qquad c = u_1 v_0 - v_1 u_0. \tag{32}$$

A reparameterization $t = t\sqrt{c^2/a^2 b^2}$ of $\mathbf{p}(t)$ given in (32) and a scaling by factor $(1/c^2)\sqrt{a^2 b^2/c^2}$ gives the curve $\mathbf{q}_2(t)$. The equality $c = 0$ obviously leads to a straight line.   □

6. Classification of Spatial Space-Like MPH Cubics

The following classification of spatial space-like MPH cubics is based on the notion of tangent indicatrix, i.e., the curve on the unit hyperboloid describing the variation of the unit tangent vector.

6.1. Orthogonal Projections into Planes Perpendicular to the Axis

Let $\mathbf{c}(t)$ be a space-like MPH cubic with curvature $\kappa \neq 0$ and torsion $\tau \neq 0$ and let $\mathbf{T}$ and $\mathbf{B}$ be the unit tangent and binormal vector of $\mathbf{c}(t)$. From the proof of Lancret's theorem (cf. Proposition 9) it follows that the direction of the axis of $\mathbf{c}(t)$ (considered as a helix in $\mathbb{R}^{2,1}$) is given by the vector $\mathbf{v} = \mathbf{T} - (\kappa/\tau)\mathbf{B}$.

One can easily verify that $\langle \mathbf{T}, \mathbf{v} \rangle = 1$ and $\|\mathbf{v}\|^2 = 1 + (\kappa^2/\tau^2)\|\mathbf{B}\|^2$. Therefore, when $\mathbf{B}$ is space-like, the vector $\mathbf{v}$ is space-like as well. In the case when $\mathbf{B}$ is time-like, the causal character of $\mathbf{v}$ may be arbitrary, e.g. the axis of $\mathbf{c}(t)$ is light-like if and only if $\|\mathbf{B}\|^2 = -1$ and $\tau = \pm\kappa$.

In the Euclidean space one can use the following approach for constructing spatial PH cubics (cubic helices). It is obvious that an orthogonal projection of a PH cubic to a plane perpendicular to its axis is a planar PH cubic (since the length of the tangent vector of the projection is a constant multiple of the length of the tangent vector of the original curve), i.e. the Tschirnhausen cubic. Therefore all PH cubics can be obtained as helices "over" the Tschirnhausen cubic by choosing its slope (or equivalently the constant ratio of its curvature to torsion). Consequently, there exists only one spatial PH cubic up to orthogonal transforms and 1D scalings in $\mathbb{R}^3$, see FAROUKI and SAKKALIS [5]. Unfortunately, in Minkowski space, this approach does not include the case of light-like axes.

Remark 16. Let $\mathbf{c}(t) = (x(t), y(t), r(t))^\top$ be a spatial space-like MPH cubic whose axis is space-like. We can suppose without loss of generality that its axis is the y axis. Then its hodograph satisfies $x'^2(t) + y'^2(t) - r'^2(t) = \sigma^2(t)$ for some polynomial $\sigma(t)$ and the orthogonal projection of $\mathbf{c}(t)$ to the xr plane is the curve $\mathbf{c}_0(t) = (x(t), 0, r(t))^\top$. As

$$|\sigma(t)| = \sqrt{x'^2(t) + y'^2(t) - r'^2(t)} = \lambda\sqrt{|x'^2(t) - r'^2(t)|} \qquad (33)$$

for some constant $\lambda \neq 0$, there exists a polynomial $\sigma_0(t)$ such that $x'^2(t) - r'^2(t) = \pm\sigma_0^2(t)$. Therefore, the orthogonal projection of a spatial space-like MPH cubic in the direction of its axis is either a planar MPH cubic or a planar *time-like* MPH cubic satisfying $x'^2(t) - r'^2(t) = -\sigma_0^2(t)$. The notion of time-like MPH curve can be generalized to space, however, there is no application for it so far. In the planar case, one can think of time-like MPH curves as of space-like MPH curves, but with swapped space and time axis. Consequently, for the remainder of the paper we include planar time-like MPH curves into planar MPH curves, as no confusion is likely to arise.

In the case of a time-like axis (the r axis) of a spatial space-like MPH cubic the orthogonal projection in the direction of its axis is a planar PH cubic, i.e. the Tschirnhausen cubic.

6.2. Classification and Normal Forms

In particular, these results concerning the tangent indicatrices apply to space-like MPH cubics (with up to two points with light-like tangents). However, in this special case of MPH curves of degree 3,

Table 1. Canonical positions of a plane π, see Fig. 5. The abbreviations sl., tl., and ll. stand for space-, time- and light-like, respectively

π	Condition	Hyperbolic angle	Canonical position	Conic section (Euclid. classif.)
sl.	$k < -l$	$\frac{1}{2}\ln\left(-\frac{k+l}{k-l}\right)$	π_s: $r = r_0$	'circle'
tl.	$k > -l, k^2 - l^2 \neq k^2$	$\frac{1}{2}\ln\left(\frac{k+l}{k-l}\right)$	π_t: $y = y_0 \neq \pm 1$	'hyperbola'
	$k > -l, k^2 - l^2 = k^2$	$\frac{1}{2}\ln\left(\frac{k+l}{k-l}\right)$	$\tilde{\pi}_t$: $y = \pm 1$	2 int. lines
ll.	$k = -l, m \neq 0$	$\ln\frac{m}{k}$	π_l: $y - r + 1 = 0$	'parabola'
	$k = -l, m = 0$	0	$\tilde{\pi}_l$: $y - r = 0$	2 par. lines

more can be achieved. As (space-like) MPH cubics are curves of a constant slope (see Proposition 9), their tangent indicatrix is a planar (conic) section of the unit hyperboloid.

Lemma 17. *Depending on its causal character, any plane π can be mapped using Lorentz transforms to one of the canonical positions shown in Table 1.*

Proof. Let us consider a plane π given by $ky + lr + m = 0$, $k \geq 0$, $l \leq 0$, $(k, l) \neq (0, 0)$, $m > 0$ (otherwise we rotate it about the time-axis and/or mirror it with respect to the xy plane). Depending on its causal character we transform π using hyperbolic rotations introduced in Table 1, cf. Fig. 5.   □

Theorem 18. *Any spatial space-like MPH cubic is similar to one of the curves listed in Table 2.*

Proof. We distinguish two cases depending on whether the axis of the MPH cubic is light-like or not.

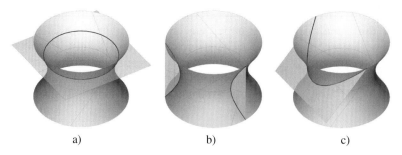

a)   b)   c)

Fig. 5. The unit hyperboloid $\mathcal{H}$, canonical positions of a plane π and corresponding conic sections (Euclidean classification): a) space-like plane π_s and a 'circle' $\mathcal{K}_s$, b) time-like plane π_t and a 'hyperbola' $\mathcal{K}_t$, c) light-like plane π_l and a 'parabola' $\mathcal{K}_l$

Table 2. Normal forms of spatial space-like MPH cubics

Axis	Normal form	1D scaling factor	κ, τ
Space-like	$(t^3 + 3t, \lambda(t^3 - 3t), 3t^2)^\top$	$\lambda \neq 0$	$\tau = -\lambda\kappa$
	$(3t^2, \lambda(t^3 - 3t), t^3 + 3t)^\top$	$\|\lambda\| > 1$	$\tau = -\lambda\kappa$
Time-like	$(3t^2, t - 3t^3, \lambda(3t^3 + t))^\top$	$\|\lambda\| < 1, \lambda \neq 0$	$\tau = \lambda\kappa$
Light-like	$\frac{1}{6}(3t^2, t^3 - 6t, t^3)^\top$		$\kappa = -\tau = 1$
	$(3t^2, t^3 - 6t, 6t)^\top$		$\tau = -\kappa$

Case 1. Let $\mathbf{p}(t)$ be a spatial space-like MPH cubic whose axis is not light-like. Its orthogonal projection in the direction of its axis is a planar MPH cubic (cf. Remark 16). Now, we make use of the classification of planar MPH cubics described in Section 5.

Subcase 1.1. The axis is time-like. We have to analyze the MPH cubics "over" the Tschirnhausen cubic $\mathbf{T}(t) = (3t^2, t - 3t^3)^\top$.

Let $\mathbf{p}(t) = (3t^2, t - 3t^3, r(t))^\top$, where $r(t) = at^3 + bt^2 + ct$. Then we have $\mathbf{p}'(t) = (6t, 1 - 9t^2, 3at^2 + 2bt + c)^\top$ and

$$\|\mathbf{p}'(t)\|^2 = (\alpha t^2 + \beta t + \gamma)^2 \qquad (34)$$

must hold for some constants α, β and γ. Comparing the coefficients in (34) gives the following system of equations:

$$\alpha^2 + 9a^2 - 81 = 0, \qquad \alpha\beta + 6ab = 0,$$
$$\beta^2 + 2\alpha\gamma + 4b^2 + 6ac - 18 = 0,$$
$$\beta\gamma + 2bc = 0, \qquad \gamma^2 + c^2 - 1 = 0. \qquad (35)$$

All solutions of the form (a, b, c) of (35) are found to be $(-3\lambda, 3\sqrt{1 - \lambda^2}, \lambda)$ and $(3\lambda, 0, \lambda)$, $|\lambda| < 1$. However, as the first family of solutions gives only planar cubics, the only spatial space-like MPH cubics with time-like axis are given by

$$\mathbf{p}(t) = (3t^2, t - 3t^3, \lambda(3t^3 + t))^\top, \qquad |\lambda| < 1, \qquad \lambda \neq 0. \qquad (36)$$

Note that there exists only one spatial space-like MPH cubic with time-like axis up to Minkowski similarities and 1D scalings given by the factor λ. For $\lambda = 0$ one obviously obtains a planar MPH cubic and the limit cases $\lambda = \pm 1$ give the W-null-cubic.

Subcase 1.2. The axis is space-like. All spatial space-like MPH cubics with space-like axes can be found analogously with the help of the planar MPH cubics presented in Proposition 15. For the sake of brevity we omit the details.

Case 2. On the other hand, let $\mathbf{q}(t)$ be a spatial space-like MPH cubic whose axis is light-like. Since the previous construction (based on Remark 16) cannot be used in this case, we turn our attention to the so-called tangent indicatrix

$$\mathbf{r}(t) = \frac{\mathbf{q}'(t)}{\|\mathbf{q}'(t)\|}. \tag{37}$$

Since the axis is light-like, the tangent indicatrix is contained in the intersection of a certain light-like plane with the unit hyperboloid $\mathcal{H}$. Without loss of generality we consider the light-like plane π_l: $y - r + 1 = 0$, cf. Table 1, Fig. 5c. (The other case of light-like planes can be omitted, since the tangent indicatrix would be contained in one of two parallel lines. This is only possible for planar curves.)

The tangent indicatrix $\mathcal{K}_l = \pi_l \cap \mathcal{H}$ has the parametric representation

$$\mathbf{r}(t) = \left(t, \frac{t^2}{2} - 1, t^2 \right)^\top.$$

All other rational biquadratic parameterizations are obtained by the bilinear reparameterizations $t = (a\tau + b)/(c\tau + d)$, $ad - bc \neq 0$ (cf. FARIN [2]) of $\mathbf{r}(t)$,

$$\tilde{\mathbf{r}}(\tau) = \left(\frac{a\tau + b}{c\tau + d}, \frac{(a\tau + b)^2 - 2(c\tau + d)^2}{2(c\tau + d)^2}, \frac{(a\tau + b)^2}{2(c\tau + d)^2} \right)^\top. \tag{38}$$

Each reparameterization leads to an MPH cubic, which is obtained by integrating the numerator, after introducing a common denominator,

$$\tilde{\mathbf{q}}(\tau) = \begin{pmatrix} \tau(2ca\tau^2 + 3\tau da + 3\tau cb + 6db) \\ \tau(a^2\tau^2 + 3a\tau b + 3b^2 - 2c^2\tau^2 - 6c\tau d - 6d^2) \\ (1/a)(a\tau + b)^3 \end{pmatrix}^\top. \tag{39}$$

Subcase 2.1. $c \neq 0$. A straightforward computation shows that we have

$$\frac{4c^4}{(bc - ad)^3} L\tilde{\mathbf{q}}\left(\frac{ad - bc}{2c^2} t - \frac{d}{c} \right) + (\tau_1, \tau_2, \tau_3)^\top = (3t^2, t^3 - 6t, 6t)^\top,$$
$$\tag{40}$$

where L is a Lorentz transform

$$L = \frac{1}{2c^2} \begin{pmatrix} 2c^2 & 2ca & -2ca \\ -2ca & 2c^2 - a^2 & a^2 \\ 2ca & a^2 & -2c^2 - a^2 \end{pmatrix} \tag{41}$$

and $(\tau_1, \tau_2, \tau_3)^\top$ is a translation vector computed from $\mathbf{q}(0) = (0,0,0)^\top$.

Subcase 2.2. $c = 0$. In this case one obtains that

$$\frac{a}{d^3} \mathbf{q}\left(\frac{d}{a}t - \frac{b}{a}\right) + (\tilde{\tau}_1, \tilde{\tau}_2, \tilde{\tau}_3)^\top = (3t^2, t^3 - 6t, t^3)^\top. \qquad (42)$$

This completes the proof. $\qquad\qquad\qquad\qquad\qquad\qquad\qquad$ □

The results are summarized in Table 2 including the ratios of curvatures to torsions of the spatial space-like MPH cubics. In the case of a space-like or time-like axis, two families or one family of space-like MPH cubics exist, respectively. In addition, there are two space-like MPH cubics with light-like axis.

7. Spatial Light-Like MPH Cubics: The *W*-Null-Cubic

We have already shown that the only spatial light-like MPH cubic is the *W*-null-cubic $\mathbf{w}(t) = (3t^2, t - 3t^3, t + 3t^3)^\top$, compare with (1) and (15). According to WUNDERLICH [13], it is the normal form of the only cubic helix (for constant slope $\alpha = \pi/4$) in Euclidean space. Moreover, as shown by WAGNER and RAVANI [11], it is also the only so-called cubic RF curve, i.e. a polynomial cubic with rational Frenet-Serret motion (of degree 5) in Euclidean space.

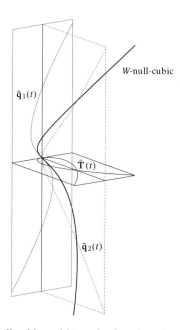

Fig. 6. The *W*-null-cubic and its projections into the coordinate planes

The orthogonal projections of the *W*-null-cubic into the *xy*, *yr* and *xr* planes are again PH or (time-like) MPH curves (see Fig. 6). The projections are similar to the Tschirnhausen cubic $\mathbf{T}(t) \in \mathbb{R}^2$ (see Section 5, Fig. 2), to the curve with one light-like tangent $\mathbf{q}_1(t) \in \mathbb{R}^{1,1}$ (cf. Proposition 15, Fig. 3) and to the curve with two light-like tangents $\mathbf{q}_2(t) \in \mathbb{R}^{1,1}$ (cf. Proposition 15, Fig. 4), respectively.

8. Conclusion

In this paper we investigated a relation between MPH cubics and cubic helices in Minkowski space. Among other results we proved that any polynomial space-like or light-like helix in $\mathbb{R}^{2,1}$ is an MPH curve. The converse result holds for cubic MPH curves, i.e. spatial MPH cubics are helices in $\mathbb{R}^{2,1}$. Based on these results and properties of tangent indicatrices of MPH curves we presented a complete classification of planar and spatial MPH cubics.

Acknowledgement

The authors thank the Austrian Science Fund (FWF) for supporting this research through project P17387-N12.

References

[1] CHOI, H. I., HAN, CH. Y., MOON, H. P., ROH, K. H., WEE, N. S. (1999) Medial axis transform and offset curves by Minkowski Pythagorean hodograph curves. Comput. Aided Des. **31**: 59–72

[2] FARIN, G. (1997) Curves and Surfaces for Computer Aided Geometric Design. Academic Press, New York

[3] FAROUKI, R. T. (2002) Pythagorean-hodograph curves. In: HOSCHEK, J., FARIN, G., KIM, M.-S. (eds.) Handbook of Computer Aided Geometric Design, pp. 405–427. Elsevier, Amsterdam

[4] FAROUKI, R. T., SAKKALIS, T. (1990) Pythagorean hodographs. IBM J. Res. Dev. **34**: 736–752

[5] FAROUKI, R. T., SAKKALIS, T. (1994) Pythagorean-hodograph space curves. Adv. Comput. Math. **2**: 41–66

[6] KIM, G.-I., AHN, M.-H. (2003) C^1 Hermite interpolation using MPH quartic. Comput. Aided Geom. Des. **20**: 469–492

[7] KOSINKA, J., JÜTTLER, B. (2006) G^1 Hermite interpolation by Minkowski Pythagorean hodograph cubics. Comput. Aided Geom. Des. **23**: 401–418

[8] KREYSZIG, E. (1991) Differential Geometry. Dover, New York

[9] KUBOTA, K. K. (1972) Pythagorean triples in unique factorization domains. Amer. Math. Month. **79**: 503–505

[10] MOON, H. P. (1999) Minkowski Pythagorean hodographs. Comput. Aided Geom. Des. **16**: 739–753

[11] WAGNER, M. G., RAVANI, B. (1997) Curves with rational Frenet-Serret motion. Comput. Aided Geom. Des. **15**: 79–101

[12] WALRAVE, J. (1995) Curves and surfaces in Minkowski space. Doctoral thesis, K. U. Leuven, Faculty of Science, Leuven

[13] WUNDERLICH, W. (1973) Algebraische Böschungslinien dritter und vierter Ordnung. Sitzungsber. Österr. Akad. Wiss., Math.-nat. Kl. Abt. II **181**: 353–376

Authors' address: Dr. Jiří Kosinka, Prof. Dr. Bert Jüttler, Institute of Applied Geometry, Johannes Kepler University, Altenberger Str. 69, 4040 Linz, Austria. E-Mail: Jiri.Kosinka@jku.at; Bert.Juettler@jku.at.

Sitzungsber. Abt. II (2006) 215: 37–44

Sitzungsberichte

Mathematisch-naturwissenschaftliche Klasse Abt. II
Mathematische, Physikalische und Technische Wissenschaften

© Österreichische Akademie der Wissenschaften 2007
Printed in Austria

Pythagoräische Tripel höherer Ordnung

Von

Edmund Hlawka

(Vorgelegt in der Sitzung der math.-nat. Klasse am 27. April 2006
durch das w. M. Edmund Hlawka)

§1.

Wir wollen das Paar (A, B) $(A, B$ ganze Zahlen, o.B.d.A. $A > B > 0)$ als Glied einer Folge

$$(A(g), B(g))$$

auffassen, wobei g alle ganzen Zahlen durchläuft.

Wir setzen zunächst

$$A(0) = A, \qquad B(0) = B, \tag{1}$$

dann definieren wir für $g \geq 0$

$$A(g + 1) = (A^2(g) + B^2(g))^{1/2} = \sqrt{S(g)},$$
$$B(g + 1) = (2A(g)B(g))^{1/2} = \sqrt{P(g)} \tag{2}$$

und für $g \leq 0$

$$A(g - 1) = \tfrac{1}{2} \left(\sqrt{S(g)} + \sqrt{D(g)} \right),$$
$$B(g - 1) = \tfrac{1}{2} \left(\sqrt{S(g)} - \sqrt{D(g)} \right). \tag{3}$$

Dabei haben wir

$$S(g) = A^2(g) + B^2(g) \tag{4}$$

und

$$D(g) = A^2(g) - B^2(g) \tag{4'}$$

gesetzt.

Ist g nichtnegativ, so erhalten wir die Rekursionsformel

$$S(g+1) = (A(g) + B(g))^2 \tag{5}$$

und für nicht positives g

$$S(g-1) = (A(g))^2. \tag{5'}$$

Analog erhalten wir für $g > 0$

$$D(g+1) = (A(g) - B(g))^2 \tag{6}$$

und für negatives g

$$D(g-1) = (A^4(g) + B^4(g))^{1/2} = \sqrt{S(g)D(g)}. \tag{6'}$$

Weiters ist für $g > 0$

$$P(g+1) = 2\sqrt{P(g)S(g)} \tag{7}$$

und für negatives g

$$P(g-1) = (B(g))^2. \tag{7'}$$

§2.

Wir definieren nun den Winkel $\vartheta(g)$ mit $\vartheta(0) = \vartheta$. Diesen Winkel habe ich in der Arbeit [3] definiert durch

$$e^{i\pi\vartheta} = \frac{A + iB}{A - iB}.$$

Er ist nach SCHERRER und HADWIGER irrational. Beschränken wir uns auf den Fall $g \geq 0$, so definieren wir

$$e^{i\pi\vartheta(g+1)} = \frac{A(g+1) + iB(g+1)}{A(g+1) - iB(g+1)} \tag{1}$$

$$= \frac{(A(g+1) - iB(g+1))^2}{A^2(g+1) + B^2(g+1)}$$

$$= \frac{A^2(g+1) - B^2(g+1) - 2iA(g+1)B(g+1)}{A^2(g+1) + B^2(g+1)}$$

$$= \frac{D(g+1) + iP(g+1)}{S(g+1)}. \tag{2}$$

Weiters definieren wir (h ganze Zahl)

$$e^{i\pi h\vartheta} = \frac{A_h(g+1) + iB_h(g+1)}{A_h(g+1) - iB_h(g+1)},$$

und es ist

$$A_h(g+1) + iB_h(g+1) = (A(g+1) + iB(g+1))^4.$$

Es ist

$$D(g+1) = (A(g) - B(g))^2$$

und

$$P(g+1) = 2\sqrt{P(g)S(g)}.$$

Es ist ($g \geq 0$)

$$\cos \pi\vartheta(g+1) = \frac{(A(g) - B(g))^2}{(A(g) + B(g))^2} = \frac{D(g)}{(A(g) + B(g))^2}$$

und

$$\sin \pi\vartheta(g+1) = \frac{2\sqrt{P(g)S(g)}}{(A(g) + B(g))^2}$$
$$= \frac{2\sqrt{2AB(A^2(g) + B^2(g))}}{A^2(g) + B^2(g) + 2A(g)B(g)},$$

also ist

$$\sin \pi\vartheta(g+1) = \frac{2\sqrt{\frac{2AB}{A^2+B^2}}}{1 + \frac{2AB}{A^2+B^2}} = \frac{2\sqrt{\sin \vartheta(g)}}{1 + \sin \vartheta(g)} \geq \sqrt{\sin \vartheta(g)}. \qquad (3)$$

Daraus folgt mit $A = A(g)$, $B = B(g)$ und $\vartheta = \vartheta(g)$

$$\sin \pi\vartheta(g+1) = \frac{2}{1 + \frac{2AB}{A^2+B^2}} = \frac{2\sqrt{\sin \vartheta}}{1 + \sin \vartheta} \geq \sqrt{\sin \vartheta}.$$

Setzen wir $\sqrt{\sin \vartheta} = y$, so ist

$$\sin \pi\vartheta(g+1) = \frac{2y}{1 + y^2}, \qquad \cos \pi\vartheta(g+1) = \frac{1 - y^2}{1 + y^2},$$

also ist

$$e^{i\pi\vartheta} = \frac{(1 - iy)^2}{(1 + y)^2} = \frac{1 - iy}{1 + iy}.$$

Es ist nun nach (3)

$$\sin \vartheta(g + 1) \geq 2\sqrt{\sin \vartheta(g)},$$

daraus folgt durch vollständige Induktion

$$\sin \vartheta(g + 1) \geq (\sin \vartheta(g))^{w(g)}$$

mit $g \geq 0$, wo $w(g) = 1/2^{g+1}$ und allgemein

$$\sin \pi h \vartheta(g + 1) \geq (\sin \pi h \vartheta(g))^{w(g)}.$$

Nun ist $\vartheta(0) = \vartheta_0$ irrational und $\sin \pi h \vartheta_0$ von der Gestalt

$$\frac{Z}{(A^2 + B^2)^h},$$

wo Z eine ganze Zahl $\neq 0$ ist.

   Es wird also

$$\sin \pi h \vartheta(g + 1) \geq \left(\frac{2A_n B_n}{A^2 + B^2}\right)^{hw(g)} \geq \frac{1}{(A^2 + B^2)^{hw(g)}}.$$

Wir betrachten nun die Weylsche Summe

$$W_h = \frac{1}{N} \sum_{h=1}^{N} e^{2\pi i h \vartheta(g+1)}.$$

Es ist nun

$$|W_h(g)| \leq \frac{1}{N} \frac{2}{|\sin h\vartheta(g + 1)|} \leq \frac{1}{N}(A^2 + B^2)^{hw(g)}.$$

Wir betrachten nun die Folge $(h\vartheta(g + 1))$ modulo 1. Nach dem Satz von ERDÖS-TURAN-KOKSMA ist die Diskrepanz dieser Folge

$$D_N \leq C\left(\frac{1}{M} + \sum_{h=1}^{M} \frac{W_N(h)}{h}\right),$$

wo M noch zu wählen und C eine absolute Konstante ist. Wir erhalten zunächst

$$D_N \leq C\left(\frac{1}{M} + \frac{1}{N}(A^2 + B^2)^{Mw(g)} \log M\right).$$

Wir wählen nun M so, dass

$$Mw(g)\lg(A^2 + B^2) - \log N = -\log \lg N$$

ist, also

$$M = \frac{\log N - \log \lg N}{w(g)\,\lg(A^2 + B^2)}$$

wird. Wir erhalten somit

$$D_N(g) \le \frac{20C\,\lg(A^2 + B^2)}{w(g)}\,\frac{\log \lg N}{\log N}.$$

Setzen wir alles ein, so erhalten wir

$$D_N(g) \le C_1 2^{g+1}\,\lg(A^2 + B^2)\left(\frac{\log \lg N}{\log N}\right).$$

Beispiel 1.

$$A(0) = 2, \qquad B(0) = 1,$$
$$A(1) = \sqrt{5}, \qquad B(1) = \sqrt{4} = 2,$$
$$A(2) = 3, \qquad \ldots$$
$$\ldots \qquad B(3) = 2\sqrt{20},$$
$$A(4) = 49, \qquad B(4) = 2\sqrt{60}.$$

Allgemein: Wir nehmen eine Primzahl von der Form $p = 4k + 1$ und definieren

$$T(s) = \sum_{x=0}^{p-1}\left(\frac{x}{p}\right)\left(\frac{x^2 + s}{p}\right),$$

wo $T(s)$ eine gerade Zahl ist. Es sei nun $s = r$, wenn r quadratischer Rest modulo p, also $(r/p) = 1$, ist. Wir schreiben $s = n$, wenn n Nichtrest modulo p, also $(n/p) = -1$, ist. p besitzt dann die Darstellung

$$p = \left(\tfrac{1}{2}T(r)\right)^2 + \left(\tfrac{1}{2}T(n)\right)^2$$

und lässt sich – wie schon von FERMAT her bekannt – als Summe zweier Quadrate darstellen. Die Darstellung ist abgesehen von Vorzeichen und Vertauschungen eindeutig. Die obige Darstellung stammt von JABOBSTHAL. Man kann also für A die Darstellung $\tfrac{1}{2}T(r)$ und für B die Darstellung $\tfrac{1}{2}T(n)$ nehmen.

§3.

Wir setzen jetzt (wieder $g \ge 0$)

$$k'(g+1) = \left(\frac{B(g+1)}{A(g+1)}\right)^2 = \frac{P(g)}{S(g)}. \tag{1}$$

Es ist

$$\sqrt{k'(g+1)} = \left(\frac{B(g+1)}{A(g+1)}\right)^2 = \sqrt{\frac{P(g)}{S(g)}} \tag{2}$$

(wir nehmen also das positive Vorzeichen der Quadratwurzel).

Wir definieren weiters, wobei wir $A(g+1)$ bzw. $B(g+1)$ kurz mit A bzw. B bezeichnen,

$$k(g+1) = \frac{1 - k'(g+1)}{1 + k'(g+1)} = \frac{A^2 - B^2}{A^2 + B^2} = \frac{D(g+1)}{S(g+1)}. \tag{3}$$

Wir erhalten also

$$k^2(g+1) + k'^2(g+1) = 1. \tag{4}$$

Es wird

$$k'(g+1) = \frac{2\frac{B(g)}{A(g)}}{1 + \frac{B^2(g)}{A^2(g)}} = \frac{2\sqrt{k'(g)}}{1 + k(g)}, \tag{5}$$

und es wird

$$1 + k(g+1) = \frac{2A^2(g)}{S(g)} = 1 + \frac{1 - k'}{1 + k'} = \frac{2}{1 + k'(g+1)}. \tag{6}$$

Wir definieren nun die Größen $a(g)$ und $b(g)$, wobei

$$a(g) = S + P, \qquad b(g) = S - P, \tag{7}$$

die wieder positiv sind, und wenden den so genannten arithmetisch-geometrischen Algorithmus (kurz AG) an. Wir setzen

$$a_1(g) = \tfrac{1}{2}(a(g) + b(g)), \qquad b_1(g) = \sqrt{a(g)b(g)}.$$

Im Fall (6) ist

$$a_1(g) = S(g), \qquad b_1(g) = \sqrt{S^2 - P^2} = D.$$

Wir setzen den Prozess weiter fort,

$$a_2(g) = \frac{a_1(g) + b_1(g)}{2}, \qquad b_2(g) = \sqrt{a_1(g)b_1(g)}$$

und so weiter.

Im zweiten Fall wenden wir AG auf das Paar

$$a'(g) = S(g) + D(g), \qquad b'(g) = S(g) - D(g) \tag{6'}$$

an. Es wird

$$a'_1(g) = S(g), \qquad b'_1(g) = \sqrt{S^2(g) - D^2(g)} = P.$$

GAUSS hat nun Folgendes gezeigt: Besteht das Ausgangspaar (a, b) aus positiven Zahlen kleiner als Eins und konstruieren wir die zugehörige AG-Folge (a_j, b_j), so sind die beiden Folgen (a_j) und (b_j) beide *konvergent* und konvergieren zum *gleichen Grenzwert* $M(a, b)$, den GAUSS den arithmetisch-geometrischen Grenzwert nennt. Es gilt weiter

$$|a_j - b_j| \leq \frac{|a - b|}{2^j}.$$

Der Grenzwert konvergiert also sehr rasch. Die Folge a_j ist monoton wachsend, die Folge b_j monoton abnehmend.

Bei der Folge (6) gilt

$$\frac{\pi}{2} \frac{1}{M(a, b)} = \int_0^{\pi/2} \frac{d\varphi}{\sqrt{a^2 \cos^2 \varphi + b^2 \sin^2 \varphi}},$$

wo $a = S$ und $b = D$ ist.

Im Fall (6′) gilt

$$\frac{\pi}{2} \frac{1}{M(a', b')} = \int_0^{\pi/2} \frac{d\varphi}{\sqrt{a'^2 \cos^2 \varphi + b'^2 \sin^2 \varphi}}$$

mit $a' = S$ und $b' = P$.

Heben wir jetzt s in k bzw. k' heraus, ersetzen also die Folge $(a(g), b(g))$ durch die Folge $(1, k(g))$ bzw. $(1, k'(g))$, so erhalten wir

$$\frac{\pi}{2} \frac{1}{M(1, k)} = \int_0^{\pi/2} \frac{d\varphi}{\sqrt{1 - k^2 \sin^2 \varphi}} = K$$

und

$$\frac{\pi}{2} \frac{1}{M(1, k')} = \int_0^{\pi/2} \frac{d\varphi}{\sqrt{1 - k'^2 \sin^2 \varphi}} = K'.$$

K und K' sind die so genannten vollständigen Integrale erster Gattung in der Legendreschen Normalform, also Funktionen von k bzw. k'.

Wir setzen weiter $q = e^{\pi i \tau}$, $\tau = iK'/K$ und führen die ϑ-Funktion ein,

$$\vartheta_3(v) = \sum q^{n^2} e^{2\pi i v},$$

wo $\vartheta_2(v) = \vartheta_1$ und $\vartheta_1(v) = \vartheta_1$ und die Nullwerte $\vartheta_j(0)$ sind. Weiters setzen wir $u = \pi\vartheta_3^2(v)$,

$$\operatorname{sn} u = \frac{1}{\sqrt{k}} \frac{\vartheta_1(v)}{\vartheta_0(v)},$$

$$\operatorname{cn} u = \sqrt{\frac{k'}{k}} \frac{\vartheta_2(v)}{\vartheta_0(v)},$$

$$\operatorname{dn} u = \sqrt{k'} \frac{\vartheta_3(v)}{\vartheta_0(v)}.$$

Es ist

$$k = \left(\frac{\vartheta_2}{\vartheta_3}\right)^2, \qquad k' = \left(\frac{\vartheta_0}{\vartheta_3}\right)^2$$

(vgl. [4], S. 53).

Literatur

[1] HLAWKA, E. (1999) Über einige geometrische Anwendungen im Zusammenhang mit Pythagoräischen Tripeln und Gleichverteilung. Aequationes Math. **58**: 163–175

[2] HLAWKA, E. (1999) Pythagorean Tripels. In: BAMBAH, R. P., DUMIR, V. C., HANS-GILL, R. J. (eds.) Number Theory, pp. 141–155. Hindustan Book Agency, New Delhi

[3] HLAWKA, E. (Preprint) Über einige geometrische Anwendungen der Pythagoräischen Tripel, Teil III

[4] KRAUSE, M. (1912) Theorie der elliptischen Funktionen (Mathematisch-physikalische Schriften für Ingenieure und Studierende, 13). B. G. Teubner, Stuttgart

Anschrift des Verfassers: Prof. Dr. Dr. h.c. Edmund Hlawka, Margarethenstraße 27/II/9, 1040 Wien, Austria.

Sitzungsber. Abt. II (2006) 215: 45–105

Sitzungsberichte
Mathematisch-naturwissenschaftliche Klasse Abt. II
Mathematische, Physikalische und Technische Wissenschaften

© Österreichische Akademie der Wissenschaften 2007
Printed in Austria

Mathematical-Physical Properties
of Musical Tone Systems II: Applications

By

Bruno J. Gruber

(Vorgelegt in der Sitzung der math.-nat. Klasse am 22. Juni 2006
durch das k. M. Peter Steinhauser)

Abstract

In [1] mathematical-physical properties of musical tones, and musical tone systems, were discussed. In the current article the results obtained in [1] are applied to analyze tonal systems in greater detail. This is done for the case of the 3-dimensional tonal lattice system, as well as for its 2-dimensional tonal sub-lattice – the Pythagorean plane.

It will be shown that in the 3-dimensional tonal lattice space a 31-tone system can be defined which contains the familiar tone scales as substructures, namely the 12-tone chromatic scales (with their subscales). Moreover, the 31-tone system contains a 3-dimensional 22/23-tone system related to a hypothetical South Indian Carnatic tone system and to the 3-tone scale of the ancient Greek Lyre.

For the case of the 2-dimensional Pythagorean sub-lattice a 29-tone system can be defined. This 29-dimensional tone lattice contains a 2-dimensional 22/23-tone system. Moreover, the 2-dimensional 29-tone system also contains the Pythagorean 12-tone systems (scales), the Pythagorean 7-tone scale and again the 3-tone scale of the ancient Greek Lyre.

Subsequently it is shown that the 2-dimensional tonal systems of the Pythagorean plane are, in fact, images of the 3-dimensional tonal systems of the 31-tonal system, if the 3-dimensional tonal systems are projected along the Pythagorean vector $p = (-1, 3, -1)$ ($= 81/80$, the Syntonic comma) into the Pythagorean plane. In particular, the chromatic minor scale projects upon the 12-tone Pythagorean scale, the chromatic major on another 12-tone Pythagorean scale (quite asymmetric with respect to the tone c), the 7-tone diatonic scale projects upon the 7-tone Pythagorean scale. The 22/23-tone 3-dimensional tone system consists of 11 pairs of tones, related to each other by the Pythagorean vector p, and the tone c. Each pair is mapped upon one tone in the Pythagorean plane, and these 11 images, together with the tone c, form the standard 12-tone Pythagorean scale. This projection establishes not only a relationship between the tones of the 3-dimensional and 2-dimensional tonal systems, but also a functional relationship between the tone intervals of various musical systems and musical scales.

Finally, it is shown that a 3-dimensional 116-tone system exists which contains all the 3-dimensional tone systems and scales mentioned above. Moreover, it contains all the musical tones (not the overtones) of the list of tones given in [2]. As an explicit example, Table 10.1 lists all the tones, the tone sequence, and the relationship between the intervals for the various tonal systems, for the first full tone T_1 (the tone interval c–d). This table also illustrates the relationship of the tones with respect to the two bases used, the (2/1), (3/2) and (5/3) basis used in this article (and in [1]), and the basis (2/1), (5/4), (3/2) basis used in [2].

The reversal of certain tone sequences of the images, due to the map along the Pythagorean vector p, is also discussed, as well as a special form for the formula for the intervals between the tones derived which expresses the intervals as a linear equation in terms of three discrete parameters.

1. Introduction

Tonal scales and tonal systems will be discussed in this article. Tonal systems, as distinguished from tonal scales, are defined as ordered inventories (Tonmaterial) of musical tones from which tones for actual musical scales can be selected, as for example the tonal system obtained by RIEMANN [2]. The ordering of the tonal inventory is however not merely an ordering according to frequency (a one-parameter ordering) but an ordering according to a lattice structure (a three-parameter ordering), of the kind used by MAZZOLA [3]. While the analysis given in this article is strictly restricted to "lattice properties" of tonal lattice systems there are obvious implications for actual musical scales [3], [4].

The musical tones are defined by ratios of frequencies ν/ν_0, whereby ν_0 is an arbitrary, but fixed reference tone. The frequencies ν within the n-th octave can be expressed in the form

$$\nu = \nu_0 2^n (1 + \delta/2\pi), \qquad 0 \leq \delta/2\pi \leq 1, \qquad n = 0, \pm 1, \pm 2, \pm 3, \ldots$$
$$(1.1a)$$

or in the form

$$\nu = \nu_0 2^{n+\xi/2\pi}, \qquad 0 \leq \xi/2\pi \leq 1, \qquad n = 0, \pm 1, \pm 2, \pm 3, \ldots$$
$$(1.1b)$$

with

$$\xi/2\pi = \log_2(1 + \delta/2\pi). \qquad (1.1c)$$

The octave tones, given by the parameter value $\delta/2\pi = 0$, are

$$\nu = \nu_0^n = \nu_0 2^n, \qquad \delta/2\pi = 0, \qquad n = 0, \pm 1, \pm 2, \pm 3, \ldots.$$
$$(1.2a)$$

The upper index n in the expression ν^n for the frequency ν denotes the n-th octave. (This notation for the n-th octave will also hold for any alphabetical musical tone like the tone a^n, in distinction to $(a)^n$ which denotes the n-th power of the frequency ratio $\nu = 5/3$ associated with the tone a). The $n = 0$ octave is called the basic octave. In this article the value chosen for ν_0 is $\nu_0 = c$, where c denotes the first tone of the diatonic scale of C-major. Thus for the case $n = 0$ holds

$$\nu_0 = \nu^0 = c. \tag{1.2b}$$

That is, the reference frequency ν_0 is chosen to be the first tone of the $n = 0$ octave which has the alphabetical name c.

Note that the expression given by Eq. (1.1) is the inverse function to the logarithmic function

$$\log_2(\nu/\nu_0) = n + \log_2(1 + \delta/2\pi) = n + \xi/2\pi. \tag{1.3}$$

Thus both expressions, Eqs. (1.1) and (1.3), carry the same information. Eq. (1.3) is essentially an expression for the distance, in cent, of the tone ν from the reference tone ν_0 (apart from a factor 1,200).

The relationship Eq. (1.1c) between frequency and perception (Hörempfinden) has also been discussed in ref. [5].

The musical tones considered in this article (and in refs. [1] and [2]) are of the form

$$\nu/\nu_0 = (n, m, r) = (2/1)^n (3/2)^m (5/3)^r,$$
$$n, m, r = 0, \pm 1, \pm 2, \pm 3 \ldots. \tag{1.4}$$

It then holds for the musical tones ν/ν_0 which are contained within the basic $(2/1)$-octave, that

$$1 \leq (2/1)^n (3/2)^m (5/3)^r = (n, m, r) \leq 2,$$
$$n, m, r = 0, \pm 1, \pm 2, \pm 3 \ldots. \tag{1.5}$$

It was shown in ref. [1] that the musical tones (n, m, r), Eq. (1.4), can be mapped onto lattice points of a 3-dimensional *scaled lattice*, with the distances between the lattice points scaled by the factor $(2/1)^n$ along the n-axis, by the factor $(3/2)^m$ along the m-axis, and by the factor $(5/3)^r$ along the r-axis. That is, each lattice point (n, m, r) represents a musical tone corresponding to the frequency ν/ν_0, as defined by Eq. (1.4).

Thus, instead of a *single number* ν/ν_0 representing a musical tone, the musical tones are now expressed in terms of *three discrete parameters*. The properties of the 3-dimensional lattice associated with the musical tones provide a mathematical structure which is not available if the musical tones are merely given in terms of pure numbers ν/ν_0

(which now represents only one property of a musical tone (n, m, r), namely "its distance" from the lattice origin). This additional mathematical structure of the musical tones (n, m, r) permits new insights into the structure of *musical tones*, and moreover supplies the mathematical tools needed to systematically study *musical tone systems*.

As it was noted above, the lattice point (vector) (n, m, r) represents a musical tone in a 3-dimensional lattice space, and with this lattice point (n, m, r) is associated a numerical value, namely the number ν/ν_0 (the frequency ratio). In this article a musical tone (n, m, r) will be understood to represent simultaneously a 3-dimensional vector, and its associated numerical value ν/ν_0. Thus, symbols like s_1, s_2, s_3, representing musical intervals, denote *simultaneously* vectors and their associated numerical value,

$$s_1 = \nu_1/\nu_0 = (n_1, m_1, r_1),$$
$$s_2 = \nu_2/\nu_0 = (n_2, m_2, r_2),$$
$$s_3 = \nu_3/\nu_0 = (n_3, m_3, r_3).$$

It is then the mathematical operation used which will distinguish between the two meanings, namely

$$s_1 + s_2 = s_3$$

is understood to be equal to the vector sum of s_1 and s_2,

$$(n_1, m_1, r_1) + (n_2, m_2, r_2) = (n_1 + n_2, m_1 + m_2, r_1 + r_2) = (n_3, m_3, r_3),$$

while the equation

$$s_1 s_2 = s_3$$

is understood to be equal to the ordinary product of the frequencies associated with s_1 and s_2,

$$(\nu_1/\nu_0)(\nu_2/\nu_0) = (\nu_3/\nu_0), \qquad \nu_0 = 1.$$

The musical tone systems, i.e. the various musical scales, form subsets of lattice points within the 3-dimensional lattice space [1]. For a set of lattice points to form a musical scale certain conditions apply. In particular the "closure condition" applies. That is, starting out with a tone $\nu_0/\nu_0 = 1 = (0, 0, 0)$ the octave tone $2\nu_0/\nu_0 = 2 = (1, 0, 0)$ must be reached in an integer number of (discrete) steps – the intervals between the tones. Thus, once the various intervals – interval vectors – have been chosen, the *lattice properties* will determine the possible musical scales which can be based upon the chosen interval vectors. The vector sum of vector-intervals, however, must not only reach the octave lattice point but must do this in such a manner that in each step only musical tones of the basic $n = 0$ octave are reached, Eqs. (1.5) and (1.6a).

Using Eq. (1.5) it is possible to determine all musical tones $\nu/\nu_0 = (n, m, r)$ which belong to the basic $n = 0$ octave $[0, 1]$. Choosing from among these musical tones certain tones as interval vectors,

$$(n_1, m_1, r_1), (n_2, m_2, r_2), (n_3, m_3, r_3), \ldots$$

it must hold

$$(0, 0, 0) = 1 \leq (0, 0, 0) + k_1(n_1, m_1, r_1) + k_2(n_2, m_2, r_2)$$
$$+ k_3(n_3, m_3, r_3) + \cdots \leq 2 = (1, 0, 0),$$
$$k_i = 0, 1, 2, \ldots, k_i^{\max}, \qquad i = 1, 2, 3, \ldots \qquad (1.6a)$$

and

$$(0, 0, 0) + k_1^{\max}(n_1, m_1, r_1) + k_2^{\max}(n_2, m_2, r_2)$$
$$+ k_3^{\max}(n_3, m_3, r_3) + \cdots = (1, 0, 0). \qquad (1.6b)$$

The number of tones N of the musical system/scale is then given by

$$N = k_1^{\max} + k_2^{\max} + k_3^{\max} + \cdots. \qquad (1.6c)$$

The formulas given by Eq. (1.6) do not determine the order of the intervals, i.e. the sequence of the tones, nor the tones themselves. The tone sequence is only partially obtained by the requirement of a monotonic increase of the interval sequence. However, since there are only two fundamental intervals s_1 and s_2 for the 2-dimensional musical scales, and since moreover the Pythagorean musical scale is assumed to form a subscale of any larger 2-dimensional musical system, the sequence for the interval-vectors s_1 and s_2 is to some extent determined. The 31-tone, 3-dimensional musical scale, is based upon three fundamental constants, namely S_1 and S_2 and $p = (-1, 3, -1) = 81/80$. Relationships obtained through the lattice properties then show that the constants S_1 and S_2 are expressible in terms of s_1, s_2 and p. This limits the possible choice of tones for 3-dimensional musical systems if the images of the 3-dimensional tones, projected along the constant vector p, are to be tones of musical systems of the 2-dimensional Pythagorean plane. The unique functional relationship between the intervals $\{s_1, s_2\}$ and $\{S_1, S_2, p\}$, given by Eqs. (6.3) and (6.4), together with the required simultaneous consistency of both musical systems within the constraints of the musical tone lattice, permits the selection of a tone system for a 31-tone 3-dimensional musical system from among the 3-dimensional tones projected along the vector p onto a 29-tone 2-dimensional tone system.

In [1] it was found that the Pythagorean musical scale is contained in a 2-dimensional sub-lattice

$$(n, m, 0) = (n, m), \qquad n, m = 0, \pm 1, \pm 2, \pm 3, \ldots \qquad (1.7)$$

of the 3-dimensional musical lattice

$$(n, m, r), \qquad n, m, r = 0, \pm 1, \pm 2, \pm 3, \ldots . \qquad (1.8)$$

The Greek Pythagorean musical scale, dating back to around 500 BCA appears to have been predated, by about 100 years, by similar musical relationships in Chinese musical theory, dating back to 600 BCA [6]. This may indicate some kind of intercultural dissemination of musical knowledge between ancient cultures, or may have been simply the result of rational reasoning inherent in human nature. This reasoning may not be limited to these two cultures but may include the South Indian Dravidian culture, whose music tradition, Carnatic music, dates back to before 1,000 BCA [7]. Using this reasoning as a work hypothesis, it may be possible that the musical traditions of these cultures may have been related, or been mutually influenced, and were possibly at some time all based on the 2-dimensional Pythagorean musical lattice system, while the 3-dimensional musical lattice system may have been a later western development. This reasoning underlies the investigation of the tonal systems of the Pythagorean(-Carnatic) plane.

Applying Eq. (1.5) to the Pythagorean(-Carnatic) plane it is found that there exist *two* tone intervals (vectors) $\{s_1, s_2\}$ such that a tonal system of 29 tones can be constructed. These two fundamental tones $\{s_1, s_2\}$ will be called srutis since they compare closely to the intervals assumed for the srutis of the Carnatic music [7]. It will then be shown that the 29-tone system contains as subsystems a 2-dimensional 23-tone and 22-tone sub-system. These two sub-systems are based upon *three* intervals $\{s_1, s_2, s_3 = (s_1 s_2)\}$, with s_3 not independent, but given in terms of s_1 and s_2. Another tonal subsystem is given by a set of 17 tones of the 29-tone tone system. These 17 tones of the 2-dimensional Pythagorean lattice are the *images* of 17 tones of the 3-dimensional tone lattice system, consisting of the tones corresponding to the 7 natural diatonic tones and the 10 tones of the sharps and flats of the chromatic scales. This subsystem is based upon the *two* intervals $\{(s_1 s_3), s_3\}$. The Pythagorean 7-tone scale is also based upon *two* intervals, namely $\{(s_1^3 s_2^2), s_3\}$, while the 3-tone scale of the ancient Greek Lyre is based upon the *two* intervals $\{(s_1^7 s_2^5), (s_1^3 s_2^2)\}$. (A parenthesis around an expression indicates that the expression acts as a unit.)

The question then arises in which manner the 2-dimensional tone systems are related to the western culture 3-dimensional tone systems. As it was mentioned above, while only two fundamental constants $\{s_1, s_2\}$ are needed for the musical scales in the Pythagorean plane, three fundamental constants $\{S_1, S_2, p\}$ are needed for the 3-dimensional musical tone systems. Moreover, it was also mentioned before, that the third constant, the vector $p = (-1, 3, -1) = 81/80$, known as the Syntonic comma, defines a map (projection) from the 3-dimensional 31-tone lattice into the 2-dimensional Pythagorean tonal plane. All tones of the 3-dimensional lattice lying on a line, defined by the vector p, are projected upon the same 2-dimensional tone. This includes, in particular also the tones and the interval factors of the standard 3-dimensional chromatic musical scales. This results in a unique functional relationship between the fundamental intervals of both systems, the properties of the 3-dimensional musical scales being reflected in the 2-dimensional musical scales, and vice versa.

The *mapping* from the 3-dimensional musical lattice space into the 2-dimensional Pythagorean lattice subspace is given by the following equation: All 3-dimensional musical lattice points

$$(n + r, m - 3r, r), \qquad r = 0, \pm 1, \pm 2, \pm 3, \ldots, \qquad (1.9)$$

are mapped, along the vector p, onto the single 2-dimensional tone

$$(n, m, 0) = (n + r, m - 3r, r) + r(-1, 3, -1),$$
$$r = 0, \pm 1, \pm 2, \pm 3, \ldots. \qquad (1.10)$$

The inverse process represents an *embedding* of a 2-dimensional lattice tone system into the 3-dimensional lattice tone system and is obviously not unique. The ambiguity for the embedding can be resolved by the requirement of simultaneous consistency of the 3-dimensional tone system and its mapped image in the Pythagorean plane. The 3-dimensional tone system can then be considered to represent a consistent embedding of the 2-dimensional tone systems into the 3-dimensional lattice space.

The 3-dimensional tonal lattice system with lattice points

$$(n, m, r), \ldots \qquad n, m, r = 0, \pm 1, \pm 2, \pm 3, \ldots, \qquad (1.11)$$

can be considered to be made up of a set of Pythagorean planes, each plane labeled by r,

$$r = 0, \pm 1, \pm 2, \pm 3, \ldots$$

such that for each fixed value r

$$(n, m, r) \qquad n, m = 0, \pm 1, \pm 2, \pm 3, \ldots \qquad (1.12)$$

represents a separate Pythagorean plane.

These Pythagorean planes can be considered to be related to each other by translations $p = (-1, 3, -1)$ such that the origin of the $r = 0$

Pythagorean plane – the tone $c = (0, 0, 0)$ – is translated into the tone corresponding to the (new) origin of the r-th Pythagorean plane – the tone $r(-1, 3, -1)$, $r = 0, \pm 1, \pm 2, \pm 3, \ldots$. These Pythagorean planes will be called equivalent with respect to the Pythagorean vector p. Similarly, musical tones related by multiples of the Pythagorean vector p will be called equivalent with respect to translations by the Pythagorean vector $p = (-1, 3, -1)$.

In what follows the names for the musical intervals/tones are taken from the "List of Intervals" given in ref. [8].

2. The Two-Dimensional Lattice Tone System

In this section the 29-tone 2-dimensional system will be derived. The tones of this system lie all in the Pythagorean plane,

$$\nu/\nu_0 = (n, m, r = 0) = (2/1)^n (3/2)^m,$$

$$n, m = 0, \pm 1, \pm 2, \pm 3, \ldots . \tag{2.1}$$

The lattice points given by Eq. (2.1) correspond to the musical lattice tones as defined in [1]. Thus the musical lattice tones form a 2-dimensional scaled sub-lattice given by

$$\nu/\nu_0 = (n, m, 0) = n(1, 0, 0) + m(0, 1, 0),$$

$$m, r = 0, \pm 1, \pm 2, \pm 3, \ldots \tag{2.2}$$

or for short,

$$\nu/\nu_0 = (n, m) = n(1, 0) + m(0, 1), \qquad m, r = 0, \pm 1, \pm 2, \pm 3, \ldots . \tag{2.3}$$

Therefore, these musical tones are uniquely characterized as lattice points (a two-parameter object)

$$(n, m), \tag{2.4}$$

having associated with them the numerical value (a one-parameter object),

$$\nu/\nu_0 = (2/1)^n (3/2)^m. \tag{2.5}$$

Eq. (2.5) represents the frequency ratio of the musical tone (n, m).

By standard convention, the numerical values ν/ν_0 for the musical tones of an octave are given alphabetic letter names like, for example, the letters c, d, e, f, g, a, $h(b)$, (c^1), for the 7 tones of the natural diatonic musical scale. In order to distinguish between the musical tones of the 3-dimensional lattice space and the subsystem of tones of the 2-dimensional Pythagorean lattice space, the tones of the latter will be characterized by a bar over the alphabetic letter,

$$\bar{c}(=c), \quad \bar{d}(=d), \quad \bar{e}, \quad \bar{f}(=f), \quad \bar{g}(=g), \quad \bar{a}, \quad \bar{h}, \quad \bar{c}^1(=c^1).$$
(2.6)

This convention is made in order that the alphabetical symbols, representing tones of the two systems which are related by the Pythagorean vector, correspond to each other, i.e. a tone with a bar, like the tone $\bar{a}$ of the Pythagorean musical scale, is the image of the 3-dimensional lattice tone without a bar, namely the image of the tone a of the natural diatonic musical scale.

The ratio of two frequency ratios corresponding to two musical tones ν_1/ν_0 and ν_2/ν_0 (the difference, if considered as vectors),

$$I_{12} = \nu_2/\nu_1,$$

is called an interval factor (interval vector) I_{12}. In order to construct a musical system it is necessary, as a first step, to find interval vectors which satisfy Eq. (1.6).

This is done by inserting, consecutively, values n, m into the inequality, Eq. (1.5),

$$1 \le (n, m, 0) = (2/1)^n (3/2)^m \le 2, \qquad n, m = 0, \pm 1, \pm 2, \pm 3, \ldots,$$
(2.7)

and then rejecting those tonal intervals $(n, m, 0)$ which do not satisfy the inequality. In this manner intervals are obtained which may be suitable for a tonal system. It is found that the two intervals

$$s_1 = (-7, 12, 0) = 3^{12}/2^{19} = 531{,}441/524{,}288 = 1.013\ 643\ 26,$$
(Pythagorean comma)

$$s_2 = (10, -17, 0) = 2^{27}/3^{17} = 134{,}217{,}728/129{,}140{,}163$$
$$= 1.039\ 318\ 25, \qquad \text{(Pyth. double diminished 3rd)} \qquad (2.8)$$

satisfy the closure condition

$$s_1^{17} s_2^{12} = 2, \qquad N = 29, \qquad (2.9)$$

for a 29-tone musical tone system. The cyclic ordering of these intervals, given by

$$\left| s_1, s_2, s_1, s_2, s_1 \right| s_1, s_2, s_1, s_2, s_1 \left| s_1, s_2, s_1, s_2, s_1 \right|$$
$$\left| s_1, s_2, s_1, s_2, s_1 \right| s_1, s_2, s_1, s_2, s_1 \left| s_1, s_2, s_1, s_2, \right. \qquad (2.10)$$

Table 2.1. The 29-Tone Two-Dimensional Musical System

$(n, m)\ \ = (2/1)^n\,(3/2)^m,\quad$ basis: $\{s_1, s_2\}$

$s_1 = (-7, 12)\ \ = 2^{-19}\,3^{12}\ \ = 1.013\ 643\ 26\ = 23.46\ \text{cent}$

$s_2 = (10, -17)\ = 2^{27}\,3^{-17}\ = 1.039\ 318\ 25\ = 66.76\ \text{cent}$

$t = s_1^3\,s_2^2 = (-1, 2)\ = 2^{-3}\,3^{2}\ = 1.125\ 000\ 00\ = 203.91\ \text{cent}$

Row $\bar{c}$:
$\bar{c} = (0,0)\ \ |s_1;\ \ \bar{d}es = (3,-5);\ s_1;\ \bar{c}is = (-4,7);\ s_2;\ \bar{z}_1 = (6,-10);\ s_1;\ \bar{d} = (-1,2)\|$
structures: $1\qquad s_1 s_2\qquad s_1^2 s_2\qquad s_1^2 s_2^2\qquad t = s_1^3 s_2^2$

Row $\bar{d}$:
$\bar{d} = (-1,2)\ \ |s_1;\ \ \bar{e}s = (2,-3);\ s_1;\ \bar{d}is = (-5,9);\ s_2;\ \bar{z}_2 = (5,-8);\ s_1;\ \bar{e} = (-2,4)\|$
structures: $s_1\qquad s_1 s_2\qquad t s_1^2 s_2\qquad t s_1^2 s_2^2\qquad t^2$

Row $\bar{e}$:
$\bar{e} = (-2,4)\ \ |s_1;\ \ \dots\ \ \bar{x}_3 = (-6,11);\ s_2;\ ges = (4,-6);\ s_1;\ \bar{f}is = (-3,6)\|$
structures: $t s_1\ \ \dots\ \ t^2 s_1^2 s_2^2\ \ t^2 s_1^2 s_2^2\ \ t^3$

Row $\bar{f}is$:
$\bar{f}is = (-3,6)\ \ |s_1;\ \ \dots\ \ \bar{x}_4 = (-7,13);\ s_2;\ \bar{a}s = (3,-4);\ s_1;\ \bar{g}is = (-4,8)\|$
structures: $t^2 s_1\ \ \dots\ \ t^3 s_1^2 s_2^2\ \ t^3 s_1^2 s_2^2\ \ t^4$

Row $\bar{g}is$:
$\bar{g}is = (-4,8)\ \ |s_1;\ \ \dots\ \ \bar{x}_5 = (-8,15);\ s_2;\ \bar{b} = (2,-2);\ s_1;\ \bar{a}is = (-5,10)\|$
structures: $t^3 s_1\ \ \dots\ \ t^4 s_1^2 s_2^2\ \ t^4 s_1^2 s_2^2\ \ t^5$

Row $\bar{a}is$:
$\bar{a}is = (-5,10)|s_1;\ \ \dots\ \ \bar{x}_6 = (-7,17);\ s_2;\ \bar{c}^1 = (1,0);\ s_1;$
structures: $t^4 s_1\ \ \dots\ \ t^5 s_1^2 s_2^2\ \ t^5 s_1^2 s_2^2 = 2$

Lower block ($\bar{y}$ / $\bar{f},\bar{g},\bar{a},\bar{h}$):

$\bar{y}_1 = (-7,12);\ s_2;\ \bar{f} = (1,-1);\ s_1;$ — structures: s_1 ; $t^2 s_1 s_2$

$\bar{y}_2 = (-8,14);\ s_2;\ \bar{g} = (0,1);\ s_1;$ — structures: $t s_1$; $t^3 s_1 s_2$

$\bar{y}_3 = (-9,16);\ s_2;\ \bar{a} = (-1,3);\ s_1;$ — structures: $t^2 s_1$; $t^4 s_1 s_2$

$\bar{y}_4 = (-10,18);\ s_2;\ \bar{h} = (-2,5);\ s_1;$ — structures: $t^3 s_1$; $t^5 s_1 s_2$

$\bar{y}_5 = (-11,20);\ s_2;$ — structure: $t^4 s_1$

$\bar{y}_6 = (-12,22);\ s_2;$ — structure: $t^5 s_1$

The interval factors/vectors are listed to the right of the tones. That is, the interval factor/vector, acting upon the tone on its left, produces the next tone at its right. The mathematical structure of a tone is indicated below the tone. It is given by the accumulation (product/sum) of interval factors/vectors

Table 2.2. The Interval Vectors and the Interval Factors for the Two-Dimensional 29-Tone Musical System

$$s_1 = \quad (-7, 12) = \quad 1.013\ 643 = 23.46\ \text{cent}$$
$$s_2 = \quad (10, -17) = \quad 1.039\ 318 = 66.76\ \text{cent}$$

The tone interval $I((n_1, m_1), (n_2, m_2))$ between two tones (n_1, m_1) and (n_2, m_2) is given, in terms of cents, by the formula, Eq. (9.9):

$$I((n_1, m_1), (n_2, m_2)) = 1{,}200(n_1 - n_2) + 701.955(m_1 - m_2)$$

$\bar{c} =$		$(0,0) =$	$2^0 3^0 =$	1.000 000
$\bar{y}_1 =$	$s_1 =$	$(-7, 12) =$	$2^{-19} 3^{12} =$	1.013 643 $(= s_1)$
$des =$	$s_1 s_2 =$	$(3, -5) =$	$2^8 3^{-5} =$	1.053 498 $(= s_3 = s_1 s_2)$
$\bar{c}is =$	$s_1^2 s_2 =$	$(-4, 7) =$	$2^{-11} 3^7 =$	1.067 871
$\bar{z}_1 =$	$s_1^2 s_2^2 =$	$(6, -10) =$	$2^{16} 3^{-10} =$	1.109 858
$\bar{d} =$	$s_1^3 s_2^2 =$	$(-1, 2) =$	$2^{-3} 3^2 =$	1.125 000
$\bar{y}_2 =$	$s_1^4 s_2^2 =$	$(-8, 14) =$	$2^{-22} 3^{14} =$	1.140 348
$\bar{e}s =$	$s_1^4 s_2^3 =$	$(2, -3) =$	$2^5 3^{-3} =$	1.185 185
$\bar{d}is =$	$s_1^5 s_2^3 =$	$(-5, 9) =$	$2^{-14} 3^9 =$	1.201 355
$\bar{z}_2 =$	$s_1^5 s_2^4 =$	$(5, -8) =$	$2^{13} 3^{-8} =$	1.248 590
$\bar{e} =$	$s_1^6 s_2^4 =$	$(-2, 4) =$	$2^{-6} 3^4 =$	1.265 625
$\bar{y}_3 =$	$s_1^7 s_2^4 =$	$(-9, 16) =$	$2^{-25} 3^{16} =$	1.282 892
$\bar{f} =$	$s_1^7 s_2^5 =$	$(1, -1) =$	$2^2 3^{-1} =$	1.333 333
$\bar{x}_3 =$	$s_1^8 s_2^5 =$	$(-6, 11) =$	$2^{-17} 3^{11} =$	1.351 524
$\bar{g}es =$	$s_1^8 s_2^6 =$	$(4, -6) =$	$2^{10} 3^{-6} =$	1.404 664
$\bar{f}is =$	$s_1^9 s_2^6 =$	$(-3, 6) =$	$2^{-9} 3^6 =$	1.423 828
$\bar{y}_4 =$	$s_1^{10} s_2^6 =$	$(-10, 18) =$	$2^{-28} 3^{18} =$	1.443 157
$\bar{g} =$	$s_1^{10} s_2^7 =$	$(0, 1) =$	$2^{-1} 3^1 =$	1.500 000
$\bar{x}_4 =$	$s_1^{11} s_2^7 =$	$(-7, 13) =$	$2^{-20} 3^{13} =$	1.520 465
$\bar{a}s =$	$s_1^{11} s_2^8 =$	$(3, -4) =$	$2^7 3^{-4} =$	1.580 025
$\bar{g}is =$	$s_1^{12} s_2^8 =$	$(-4, 8) =$	$2^{-12} 3^8 =$	1.601 807
$\bar{y}_5 =$	$s_1^{13} s_2^8 =$	$(-11, 20) =$	$2^{-31} 3^{20} =$	1.623 660
$\bar{a} =$	$s_1^{13} s_2^9 =$	$(-1, 3) =$	$2^{-4} 3^3 =$	1.687 500
$\bar{x}_5 =$	$s_1^{14} s_2^9 =$	$(-8, 15) =$	$2^{-23} 3^{15} =$	1.710 523
$\bar{b} =$	$s_1^{14} s_2^{10} =$	$(2, -2) =$	$2^4 3^{-2} =$	1.777 777
$\bar{a}is =$	$s_1^{15} s_2^{10} =$	$(-5, 10) =$	$2^{-15} 3^{10} =$	1.802 032
$\bar{y}_6 =$	$s_1^{16} s_2^{10} =$	$(-12, 22) =$	$2^{-34} 3^{22} =$	1.862 618
$\bar{h} =$	$s_1^{16} s_2^{11} =$	$(-2, 5) =$	$2^{-7} 3^5 =$	1.898 437
$\bar{x}_6 =$	$s_1^{17} s_2^{11} =$	$(-9, 17) =$	$2^{-26} 3^{17} =$	1.924 338
$\bar{c}^1 =$	$s_1^{17} s_2^{12} =$	$(1, 0) =$	$2^1 3^0 =$	2.000 000

yields a musical system. Calculation of the cent (out of 1,200 cent) yields the values

$$s_1 = 23.46 \text{ cent},$$
$$s_2 = 66.76 \text{ cent}. \tag{2.11}$$

A third interval factor, to be of importance later on, is given by

$$s_3 = s_1 s_2 \ (= s_1 + s_2, \text{ in vector form}) = 90.22 \text{ cent (Limma)}. \tag{2.12}$$

These values correspond closely to the approximate values of the sruti (interval factors/vectors) for a hypothetical 22-tone (or 23-tone) Carnatic musical system, given in [7], namely

$$s_1 = 22 \text{ cent}, \qquad s_2 = 66 \text{ cent}, \qquad \text{and} \qquad s_3 = 90 \text{ cent}.$$

Thus the three interval factors s_1, s_2, s_3 will be referred to as sruti.

It is then seen that a 29-tone musical system is obtained, with the closure condition (2.9), and is given in vector form, by

$$17(-7, 12) + 12(10, -17) = (1, 0) = \bar{c}^1 = c^1 \tag{2.13a}$$

or equivalently by

$$s_1^{17} s_2^{12} = 2. \tag{2.13b}$$

This implies that the tone c_0 $(= c^0$, with upper *index* $n = 0)$ is "musically equivalent" to the tone c^1 (though not identical to c^1), but differing from c_0 by the scaling factor 2, with the next octave "cycle" starting with the tone c^1.

Thus there exists a cycle of 29 tonal steps (a sequence of sruti s_1 and s_2) such that the sequence "closes" (i.e. the tone c^1 is reached, the "octet condition"). Moreover it holds that

$$3(-7, 12) + 2(10, -17) = (-1, 2) = \bar{d} = d,$$
$$s_1^3 s_2^2 = 9/8,$$
$$2(3(-7, 12) + 2(10, -17)) = (-2, 4) = \bar{e} = e + p,$$
$$\left(s_1^3 s_2^2\right)^2 = 81/64 = (5/4)(81/80) = e \cdot p,$$
$$3(3(-7, 12) + 2(10, -17)) = (-3, 6) = \bar{fis} = fis + 2p,$$
$$\left(s_1^3 s_2^2\right)^3 = 729/512 = (25/18)(81/80)^2 = (fis) \cdot p^2,$$
$$4(3(-7, 12) + 2(10, -17)) = (-4, 8) = \bar{gis} = gis + 2p,$$
$$\left(s_1^3 s_2^2\right)^4 = 6,561/4,096 = (25/16)(81/80)^2 = (gis) \cdot p^2,$$

$$5(3(-7, 12) + 2(10, -17)) = (-5, 10) = \bar{a}is = ais + 3p,$$

$$(s_1^3 s_2^2)^5 = 59{,}049/32{,}768 = (125/72)(81/80)^3 = (ais) \cdot p^3,$$

$$5(3(-7, 12) + 2(10, -17)) + 2(-7, 12) + 2(10, -17)$$
$$= (1, 0, 0) = \bar{c}^1 = c^1,$$

$$(s_1^3 s_2^2)^5 (s_1^2 s_2^2) = 2 = \bar{c}^1 = c^1. \tag{2.14}$$

The 29-tone 2-dimensional musical system is given in Table 2.1. The tones $\bar{x}, \bar{y}, \bar{z}, \bar{w}$ listed in this table appear not to have standardized names. It will be noted that this table exhibits a great amount of symmetry. Table 2.2 lists the numerical values for the tones of the 29-tone musical system.

3. The Two-Dimensional 23/22 Musical Tone Systems

A 22/23-tone system can be obtained from the 29-tone system, discussed above, by choosing three distinct intervals, namely three sruti. Choosing the two independent sruti, s_1 and s_2, and forming a third sruti, $s_3 = (s_1 s_2)$, the set of three sruti

$$\{s_1, s_2, s_3 = (s_1 s_2)\} \tag{3.1}$$

forms the intervals for the 22/23-tone musical system.

It might be remarked that, if the Carnatic musical scale should have been a Pythagorean type scale, the uncertainty concerning the actual number of tones of the Carnatic musical scale, 22 tones or more [7], or denying the existence of a Carnatic musical scale [9], may have to do with the fact that out of the 29-tone 2-dimensional musical system at different times different numbers of tones were selected to form different Carnatic tone systems.

The sequence of intervals for the Carnatic scales/systems is given by

$$\left| s_3, s_1, s_2, s_1 \right| s_3, s_1, s_2, s_1 \left| s_3, s_1, s_2, s_1 \right|$$
$$\left| s_3, s_1, s_2, s_1 \right| s_3, s_1, s_2, s_1 \left| s_3, s_1, s_2, \right. \qquad N = 23 \text{ tones},$$

or

$$\left| s_3, s_1, s_2, s_1 \right| s_3, s_1, s_2, s_1 \left| s_3, s_1, s_2, s_1 \right|$$
$$\left| s_3, s_1, s_2, s_1 \right| s_3, s_1, s_2, s_1 \left| s_3, s_3, \right. \qquad N = 22 \text{ tones}$$

$$t = s_3 \, s_1^2 \, s_2 = (-1, 2) = 9/8 = \bar{d} = d. \tag{3.2}$$

It will be noted that the sequence of sharps and flats in the Carnatic scale, like $\overline{des}$ and $\overline{cis}$, is reversed from the sequence of sharp and flats des and cis, in the chromatic scales. This is caused by the prop-

Table 3.1. The Two-Dimensional Carnatic 23-Tone and 22-Tone Scales (Systems)

(n,m)	$= (2/1)^n (3/2)^m$		
$s_1 = (-7, 12)$	$= 2^{-19}3^{12}$	$= 1.013\,643\,26$	$= 23.46$ cent
$s_2 = (10, -17)$	$= 2^{27}3^{-17}$	$= 1.039\,318\,25$	$= 66.76$ cent
$t = s_1^3 s_2^2 = (-1, 2)$	$= 2^{-3}3^2$	$= 1.125\,000\,00$	$= 203.91$ cent

Tone		Tone		Tone		Tone		Tone	
$\bar{c} = (0,0)$ |s_3;	1	$\overline{des} = (3,-5)$; s_1;	s_3	$\bar{c}is = (-4,7)$; s_2;	$s_3 s_1$	$\bar{z}_1 = (6,-10)$; s_1;	$s_3 s_1 s_2$	$\bar{d} = (-1,2)\|$	$t = s_3 s_1^2 s_2$
$\bar{d} = (-1,2)$ |s_3;		$\bar{e}s = (2,-3)$; s_1;	ts_3	$\bar{d}is = (-5,9)$; s_2;	$ts_3 s_1$	$\bar{z}_2 = (5,-8)$; s_1;	$ts_3 s_1 s_2$	$\bar{e} = (-2,4)\|$	t^2
$\bar{e} = (-2,4)$ |s_3;		$\bar{f} = (1,-1)$; s_1;	$t^2 s_3$	$\bar{x}_3 = (-6,11)$; s_2;	$t^2 s_3 s_1$	$\overline{ges} = (4,-6)$; s_1;	$t^2 s_3 s_1 s_2$	$\bar{f}is = (-3,6)\|$	t^3
$\bar{f}is = (-3,6)$ |s_3;		$\bar{g} = (0,1)$; s_1;	$t^3 s_3$	$\bar{x}_4 = (-7,13)$; s_2;	$t^3 s_3 s_1$	$\bar{a}s = (3,-4)$; s_1;	$t^3 s_3 s_1 s_2$	$\bar{g}is = (-4,8)\|$	t^4
$\bar{g}is = (-4,8)$ |s_3;		$\bar{a} = (-1,3)$; s_1;	$t^4 s_3$	$\bar{x}_5 = (-8,15)$; s_2;	$t^4 s_3 s_1$	$\bar{b} = (2,-2)$; s_1;	$t^4 s_3 s_1 s_2$	$\bar{a}is = (-5,10)\|$	t^5
$\bar{a}is = (-5,10)$ |s_3;		$\bar{h} = (-2,5)$; s_1;	$t^5 s_3$	$\bar{x}_6 = (-7,17)$; s_2;	$t^5 s_3 s_1$	$\bar{c}^1 = (1,0)$;	$t^5 s_3 s_1 s_2 = 2 = t^6 s_1^{-1}$	**N = 23 tones**	
$\bar{a}is = (-5,10)$ |s_3;		$\bar{h} = (-2,5)$; s_3;	$t^5 s_3$			$\bar{c}^1 = (1,0)$;	$t^5 s_3^2 = 2 = t^6 s_1^{-1}$	**N = 22 tones**	

The interval factors/vectors are listed to the right of the tones. That is, the interval factor/vector, acting upon the tone on its left, produces the next tone at its right. The mathematical structure of a tone is indicated below the tone. It is given by the accumulation (product/sum) of interval factors/vectors erty of the map, along the Pythagorean vector $p = (-1, 3, -1) = 81/80$, from the (scaled) chromatic lattice into the (scaled) Pythagorean plane (underlying nonlinear properties of the tonal lattice). For details on the structure of tones and the intervals see Table 3.1.

4. Scales Contained in the Two-Dimensional 23/22 Musical Tone System

In this section some of the musical systems/scales will be discussed which can be derived from the Carnatic musical system. The scales/systems discussed are

(1) A 17-tone system consisting of the images in the Pythagorean plane of the set of the 7 natural diatonic tones, the 5 sharps and the 5 flats, see Sect. 7;
(2) two 12-tone Pythagorean scales;
(3) the Pythagorean musical 7-tone scale, and
(4) the 3-tone scale of the ancient Greek Lyre.

A summary for these musical scales can be found in Table 4.1.

(1) The 17-tone system is obtained from the 22-tone Carnatic system/scale by choosing the two interval factors as

$$\{s_1, s_3 = (s_1 s_2)\}$$

and choosing the interval sequence

$$|s_3, s_1, s_3|s_3, s_1, s_3|s_3, s_3, s_1|$$
$$|s_3, s_3, s_1|s_3, s_3, s_1|s_3, s_3, \qquad N = 17 \text{ tones.} \qquad (4.1)$$

Thus this system depends on the two sruti s_1 and s_3 only.

(2) The Pythagorean 12-tone scales are based upon the interval factors

$$\{(s_3 s_1), s_3\},$$

with the interval sequences

$$|(s_3 s_1), s_3|(s_3 s_1), s_3|s_3/(s_3 s_1)|s_3/(s_3 s_1)|s_3/(s_3 s_1)|s_3/s_3$$
$$|(s_3 s_1), s_3|(s_3 s_1), s_3|s_3/s_3(s_1|s_3)/s_3(s_1|s_3)/s_3(s_1|s_3)/s_3$$
$$(s_3 s_1)^5 (s_3)^7 = 2, \qquad N = 12.$$

(3) The Pythagorean 7-tone scale proper is obtained by choosing the two interval factors

$$\{t = s_1^3 s_2^2, \ s = s_3\},$$

Table 4.1. Summary of the Two-Dimensional 29-Tone Musical System

$\{s_1 = (-7,12) = 2^{-19}3^{12}, \quad s_2 = (10,-17) = 2^{27}3^{-17}\}, \quad \{t = (\bar{s}_1^3\,\bar{s}_2^2) = (s_1 s_2^2), \quad s = s_3 = s_1 s_2\}, \quad c^1 = 2 = (1,0,0)$

The 29-tone system, $\{s_1, s_2\}$, $N = 29$:

$c|s_1\ s_2\ s_2\ s_1\ |s_1\ s_2\ s_2\ s_1\ s_2\ s_1\ |s_1s_2/s_1\ s_2\ s_2\ s_1|s_1\ s_2/s_1\ s_2\ s_1|s_1\ s_2/s_1\ s_2; \quad c^1 = s_1^{17}\ s_2^{12}.$

Carnatic 23-tone system/scale, $s_3 = s_1 + s_2 = (3,-5) = s_1\ s_2$ $\{s_1, s_2, s_3\}, \quad N = 23:$

$c|s_3\ s_1\ s_2\ s_1|s_3\ s_1\ s_2\ s_1|s_3/s_1\ s_2\ s_1|s_3/s_1\ s_2\ s_1|s_3/s_1\ s_2\ s_2; \quad c^1 = s_1^{11}\ s_2^6\ s_3^6.$

Carnatic 22-tone systems/scales, $\{s_1, s_2, s_3\}, \quad N = 22:$

$c|s_3\ s_1\ s_2\ s_1|s_3\ s_1\ s_2\ s_1|s_3/s_1\ s_2\ s_1|s_3/s_1\ s_2\ s_1|s_3/s_1\ s_2\ s_1|s_3/s_3; \quad c^1 = s_1^{10}\ s_2^5\ s_3^7;$

$c|s_3\ s_1\ s_2\ s_1|s_3\ s_1\ s_2\ s_1|s_3/s_1\ s_2\ s_1|s_1\ s_2/s_1\ s_3|s_2\ s_2\ s_1/s_3\ s_1|s_2\ s_2\ s_1/s_3; \quad c^1 = s_1^{10}\ s_2^5\ s_3^7.$

Carnatic 17-tone system, $\{s_1, s_3\}, \quad N = 17:$

$c|s_3\ s_1\ s_3|s_3\ s_1\ s_3|s_3/s_3\ s_1|s_3/s_3\ s_1|s_3/s_3\ s_1|s_3/s_3; \quad c^1 = s_1^5\ s_3^{12}.$

"Carnatic major 12-tone scale", $\{(s_3s_1), s_3\}, \quad \{t = (\bar{s}_1^3\bar{s}_2^2), s = s_3\}, \quad N = 12:$

$c|(s_3s_1)\ s_3|(s_3s_1)\ s_3|(s_3s_1)\ s_3/(s_3s_1)|s_3/(s_3s_1)\ s_3/(s_3s_1)|s_3/s_3; \quad c^1 = (s_3s_1)^5\ s_3^7;$

$c|(t\ s^{-1})\ s|(t\ s^{-1})\ s|s/(t\ s^{-1})\ s|s/(t\ s^{-1})|s/s; \quad c^1 = (t\ s^{-1})^5\ s^7.$

"Carnatic minor" 12-tone scale – Pythagorean 12-tone scale, $\{s_3, (s_1s_3)\}, \quad \{t = (\bar{s}_1^3\bar{s}_2^2), s = s_3\}, \quad N = 12:$

$c|s_3\ (s_1s_3)|s_3\ s_3\ (s_1s_3)|s_3\ s_3\ (s_1s_3)/s_3\ (s_1s_3)/s_3\ (s_1s_3)/s_3; \quad c^1 = (s_3s_1)^5\ s_3^7;$

$c|s(t\ s^{-1})|s(t\ s^{-1})|s/s(t|s^{-1})/s(t|s^{-1})/s(t|s^{-1})/s; \quad c^1 = (t\ s^{-1})^5\ s^7.$

Pythagorean 7-tone scale, $\{t = (\bar{s}_1^3\bar{s}_2^2), s = s_3\}, \quad N = 7:$

$c|(s_1s_3^2)|(s_1s_3^2)|s_3/(s_1s_3^2)/(s_1|s_3^2)/s_3; \quad c^1 = (s_1s_3^2)^5\ s_3^2; 7$

$c|t|t|s/t/t/s;$ $c^1 = t^5\ s^2; 7.$

Ancient Greek Lyre 3-tone scale, $\{((s_1s_3^2)^2\ s_3), (s_1s_3^2)\}, \quad N = 3:$

$c|((s_1s_3^2)^2\ s_3)/((s_1s_3^2)^2\ s_3); \quad c^1 = ((s_1s_3^2)^2\ (s_1s_3^2);$

$c|((t^2\ s)^2\ s)/((t^2\ s)^2\ s); \quad c^1 = ((t^2\ s)^2\ s)^2\ (t^2\ s).$

with the interval sequence

$$|t|t|s/t/t/t/s, \quad t^5 s^2 = (s_1^3 s_2^2)^5 s_3^2 = 2, \qquad N = 7. \qquad (4.2)$$

(4) The two intervals for the ancient Greek Lyre are given by

$$\{t^2 s = (s_1^7 \, s_2^5), t = (s_1^3 \, s_2^2)\}, \qquad t = s_1^3 \, s_2^2, \qquad s = s_3$$

with

$$(t^2 s)^2 t = 2, \qquad N = 3.$$

The tone scale is given by the three tones

$$c(=\bar{c}); t^2 s \rightarrow \bar{f}(=f); t \rightarrow \bar{g}(=g); t^2 s \rightarrow \bar{c}^1(=c). \qquad (4.3)$$

The tones of the ancient Greek Lyre musical scale are simultaneously tones of the 3-dimensional and 2-dimensional musical systems.

Table 4.1 gives a summary of the properties of the Pythagorean plane-based tonal systems in terms of the two sruti s_1 and s_2.

Whether, or not, the hypothetical Carnatic musical scales derived in this article were ever used in practice is disputed, ref. [9]. However, the mathematical system for musical systems/scales developed in this article does, in a natural way, lead to tonal systems/scales of 29, 23, 22, 12, 7 and 3 tones, numbers which either correspond to established scales or keep coming up in discussions among scholars concerning the Carnatic scales, refs. [10], [11]. That a mathematical theory predicts precisely these numbers – and not other numbers – appears to be beyond a mere, unrelated, coincidence. In addition, in subsequent sections of this article it will be shown that the 29-tone 2-dimensional musical system developed in this article is functionally correlated to a 31-tone 3-dimensional lattice system, which in turn contains the standard chromatic musical scales as subscales. The mathematical structures and correlations discussed in this article appear to reflect themselves in theoretical discussions, as well as in practical constructions, of musical scales by musicians.

5. Comments on the Embedding of Musical Scales and Tone Systems

In this section the relationship between the *tones* of the *tone systems* in the *3-dimensional lattice space* and the *tones* of the *tone systems* in the *2-dimensional Pythagorean sub-lattice* is discussed. This is in view of embedding the 2-dimensional Pythagorean musical systems/scales into the 3-dimensional musical lattice space.

An embedding is given if, for each tone of a 2-dimensional Pythagorean tone system, a tone in 3-dimensional lattice space can be found, such that the collection of these 3-dimensional lattice tones satisfies

(a) all the properties of a 3-dimensional lattice tone system, and
(b) the images of the 3-dimensional tones in the Pythagorean lattice space, and the images of the properties of the 3-dimensional tone system in the Pythagorean plane become the tones and the properties of the Pythagorean tone system.

The Pythagorean tone system is then said to be *embedded* in the 3-dimensional lattice tone system.

The Pythagorean lattice space with lattice points

$$(n, m, 0), \qquad n, m = 0, \pm1, \pm2, \pm3, \ldots \qquad (5.1)$$

forms a sub-lattice of the 3-dimensional lattice space with lattice points

$$(n, m, r), \qquad n, m, r = 0, \pm1, \pm2, \pm3, \ldots \qquad (5.2)$$

for $r = 0$.

The tones

$$c = \bar{c} = (0,0,0), \quad d = \bar{d} = (-1, 2, 0), \quad f = \bar{f} = (1, -1, 0),$$
$$1 \qquad\qquad\qquad 9/8 \qquad\qquad\qquad 4/3$$
$$g = \bar{g} = (0, 1, 0), \quad c^1 = \bar{c}^1 = (1, 0, 0), \qquad\qquad\qquad (5.3)$$
$$3/2 \qquad\qquad\qquad 2$$

are simultaneously tones of both, the 3-dimensional tone system and the Pythagorean tone system, while the tones

$$e = (-1, 1, 1), \quad a = (0, 0, 1), \quad h(b) = (-1, 2, 1), \qquad (5.4)$$
$$5/4 \qquad\qquad\qquad 5/3 \qquad\qquad\qquad 15/8$$

are tones of the 3-dimensional lattice tone system, but not of the Pythagorean tone system. These tones however can be mapped onto tones of the Pythagorean-Carnatic plane by means of the projection

$$(n, m, r) + r(-1, 3, -1) = (n - r, m + 3r, 0), \qquad (5.5)$$

where the vector

$$p = (-1, 3, -1) = (2/1)^{-1}(3/2)^3(5/3)^{-1} = 81/80 \qquad (5.6)$$

has the numerical value of the syntonic comma. The 3-dimensional tones e, a, h, Eq. (5.4), are then related to their images $\bar{e}$, $\bar{a}$, $\bar{h}$ in the Pythagorean plane by

$$e + p = (-2, 4, 0) = \bar{e}, \qquad ep = (5/4)(81/80) = 81/64 = \bar{e},$$
$$a + p = (-1, 3, 0) = \bar{a}, \qquad ap = (5/3)(81/80) = 27/16 = \bar{a},$$
$$h + p = (-2, 5, 0) = \bar{h}, \qquad hp = (15/8)(81/80) = (243/128) = \bar{h}.$$

$$(5.7)$$

The projection along the vector p into the Pythagorean plane is however, as it was pointed out before, not one to one. The problem is then to identify, from among all the 3-dimensional lattice tones which are projected upon a given 2-dimensional tone, in a unique way that particular 3-dimensional tone which also satisfies the properties of a (possibly) larger 3-dimensional lattice tone system. In other words, a (possibly) larger 3-dimensional lattice tone system needs to be found such that the properties of the 3-dimensional lattice tone system, projected into the 2-dimensional Pythagorean lattice space, yields the Pythagorean musical system, while the properties of both systems must be simultaneously satisfied.

Such a situation arises for the case of the 29-tone 2-dimensional Pythagorean tone system. It will be shown that the 29-tone system can be embedded into the 3-dimensional lattice tone space, such that 31 3-dimensional lattice tones can be determined which form a 3-dimensional musical lattice tone system. Of the 31 3-dimensional lattice tones, 29 are projected onto 29 2-dimensional lattice tones of the basic $n = 0$ octave. The images of the remaining 2 tones, namely the tones $\bar{y}_1$ and $\bar{w}_1$, are in the $n = -1$ and the $n = 1$ octaves of the 2-dimensional lattice tone system and are thus "lost tones". That is, these two tones are *not* tones of the 29-tone tone system. Also a *reversal* of tone sequence occurs, see Tables 6.2 and 6.3.

Another example for this relationship of mapping and embedding is the map of the hypothetical 3-dimensional 23-tone Carnatic system/scale into the Pythagorean plane resulting into the 12-tone Pythagorean musical scale. The 12-tone Pythagorean scale is, vice versa, embedded into the larger 23-tone 3-dimensional system such that, except for the tone c, two 3-dimensional Carnatic tones correspond to one 2-dimensional Pythagorean tone and the properties of both systems are simultaneously satisfied. This then implies a well defined relationship for the intervals of the two tonal systems. See Sect. 10 and ref. [10].

6. The Three-Dimensional 31-Tone Musical Tone System

It was shown in [1] that the interval factors for the standard chromatic tone systems can be expressed in terms of the three constant vectors (intervals)

$$T_2 = S_2 = (1, 1, -2) = 27/25 = 1.080\ 000\ 00,$$
$$T_1 = S_3 = (-2, 1, 2) = 25/24 = 1.041\ 666\ 67,$$
$$S = (2, -2, -1) = (1, 1, -2) - (-1, 3, -1) = S_2 p^{-1}$$
$$= 16/15 = 1.066\ 666\ 67. \tag{6.1}$$

The symbols T_1 and T_2 denote the two (distinct) tones of the 3-dimensional musical system, while the symbol S denotes its semitone. The symbols S_2 and S_3 have been introduced for reasons to become clear later on.

It was pointed out in the previous section that the two tones c and d of the chromatic musical scale are also tones of the Pythagorean musical scale, $c = \bar{c}$, $d = \bar{d}$. That is, these two tones are lattice points of the Pythagorean plane and satisfy the properties required by both, the chromatic tone scale and the Pythagorean tone scale,

$$
\begin{array}{ccc}
(0,0) & \xrightarrow{t=(-1,2)} & (-1,2) \\
\bar{c} = c & & \bar{d} = d \\
(0,0,0) \xrightarrow{T_1=(-2,1,2)} & (-2,1,2) \xrightarrow{T_2=(1,1,-2)} & (-1,2,0), \quad (6.2) \\
c & cis & d
\end{array}
$$

where $t = (-1, 2) = 9/8$ denotes an interval factor of the Pythagorean musical scale.

The question then arises whether additional intervals can be introduced which give rise to tones lying in between the tones c and d (and thus also in between the other tones of the chromatic scale) such that (a) an "enlarged chromatic system/scale" can be constructed, and (b) the projection of the "enlarged chromatic system", along the Pythagorean vector $p = (-1, 3, -1)$ into the Pythagorean plane, results into the 29-tone 2-dimensional Pythagorean musical system. The 31-tone 3-dimensional system/scale will then contain all the standard chromatic scales and subscales, and by means of projection also the 29-tone 2-dimensional system and its subsystems. This will be demonstrated in what follows.

Table 6.1. The Three-Dimensional 31 Tone System

$$\{S_1^{-1} = (3, 0, -4),\ S_1^2 S_2 = (-5, 1, 6),\ S_1^{-1} \cdot p^{-1} = (4, -3, -3)\};\quad (S_1^{-1})^{15}(S_1^2 S_2)^{12}(S_1^{-1}p^{-1})^4 = 2$$

c| (0,0,0); S_1^{-1};	*y*$_1$ (3,0,−4); $S_1^2 S_2$;	*cis* (−2,1,2); S_1^{-1};	*des* (1,1,−2); $S_1^2 S_2$;	*y*$_2$ (−4,2,4); S_1^{-1};	*d*|| (−1,2,0)
d| S_1^{-1};	*z*$_1$ (2,2,−4); $S_1^2 S_2$;	*dis* (−3,3,2); $S_1^{-1}p^{-1}$;	*es* (1,0,−1); $S_1^2 S_2$;	*y*$_3$ (−4,1,5); S_1^{-1};	*e*| (−1,1,1)
e| $S_1^{-1}p^{-1}$;	*z*$_2$ (3,−2,−2); $S_1^2 S_2$;	*x*$_3$ (−2,−1,4); S_1^{-1};	*f* (1,−1,0); $S_1^2 S_2$;	*y*$_4$ (−4,0,6); S_1^{-1};	*fis*| (−1,0,2)
fis| S_1^{-1};	*ges* (2,0,−2); $S_1^2 S_2$;	*x*$_4$ (−3,1,4); S_1^{-1};	*g* (0,1,0); $S_1^2 S_2$;	*y*$_5$ (−5,2,6); S_1^{-1};	*gis*| (−2,2,2)
gis| $S_1^{-1}p^{-1}$;	*as* (2,−1,−1); $S_1^2 S_2$;	*x*$_5$ (−3,0,5); S_1^{-1};	*a* (0,0,1); $S_1^2 S_2$;	*y*$_6$ (−5,2,6); S_1^{-1};	*ais*| (−2,1,3)
ais| S_1^{-1};	*b* (1,1,−1); $S_1^2 S_2$;	*x*$_6$ (−4,2,5); S_1^{-1};	*h* (−1,2,1); $S_1^2 S_2$;	*w*$_1$ (−6,3,7); S_1^{-1};	*w*$_2$| (−3,3,3)
w$_2$| $S_1^{-1}p^{-1}$;	*c*^{1} (1,0,0);				

The interval factors/vectors are listed to the right of the tones. That is, the interval factor/vector, acting upon the tone on its left, produces the next tone at its right

Table 6.2. Correlation Between the Three-Dimensional 31-Tone System and the Two-Dimensional 29-Tone Systems

$S_1^{-1} = (3,0,-4) = 1.036\ 800\ 00,\qquad s_1 = (-7,12,0) = S_1 p^4 = 1.013\ 643\ 26$

$S_1^2 S_2 = (-5,1,6) = 1.004\ 693\ 93,\qquad s_2 = (10,-17,0) = S_1^{-1} S_2 p^{-6} = 1.039\ 318\ 25$

$S_1^{-1} p^{-1} = (4,-3,-3) = 1.024\ 000\ 00,\qquad s_3 = (-3,5,0) = s_1 s_2$

$S_1^{-1} S_2 = (4,1,-6),\qquad p = (-1,3,-1)$

The tone interval $I((n_1,m_1,r_1),(n_2,m_2,r_2))$ between two tones (n_1,m_1,r_1) and (n_2,m_2,r_2) is given, in terms of cents, by the formula Eq. (9.9):

$$I((n_1,m_1,r_1),(n_2,m_2,r_2)) = 1{,}200(n_1-n_2) + 701.955(m_1-m_2) + 884.358\ 71(r_1-r_2)$$

$w_2^{-1} = S_1 p =$	$(-4,3,3)$	$= 0.976\ 562\ 56 < 1$	$\bar{y}_1 = (7,-12,0) = y_1 - 4p =$	$0.986\ 540\ 56 < 1$
$c =$	$(0,0,0)$	$= 1.000\ 000\ 00$	$\bar{c} = (0,0,0) = c =$	$1.000\ 000$
$y_1 =$	$(3,0,-4)$	$= 1.036\ 800\ 00$	$\bar{w}_2^{-1} = (-7,12,0) = w_2^{-1} + 4p =$	$1.013\ 643$
$cis =$	$(-2,1,2)$	$= 1.041\ 666\ 67$	$\bar{des} = (3,-5,0) = des - 2p =$	$1.053\ 498$
$des =$	$(1,1,-2)$	$= 1.080\ 000\ 00$	$\bar{cis} = (-4,7,0) = cis + 2p =$	$1.067\ 871$
$y_2 =$	$(-4,2,4)$	$= 1.085\ 069\ 00$	$\bar{z}_1 = (6,-10,0) = z_1 - 4p =$	$1.109\ 858$
$d =$	$(-1,2,0)$	$= 1.125\ 000\ 00$	$\bar{d} = (-1,2,0) = d$	
$z_1 =$	$(2,2,-4)$	$= 1.166\ 400\ 00$	$\bar{y}_2 = (-8,14,0) = y_2 + 4p =$	$1.140\ 348$
$dis =$	$(-3,3,2)$	$= 1.171\ 875\ 00$	$\bar{es} = (2,-3,0) = es - p =$	$1.185\ 185$
$es =$	$(1,0,-1)$	$= 1.200\ 000\ 00$	$\bar{dis} = (-5,9,0) = dis + 2p =$	$1.201\ 355$
$y_3 =$	$(-4,1,5)$	$= 1.205\ 632\ 72$	$\bar{z}_2 = (5,-8,0) = z_2 - 2p =$	$1.248\ 590$
$e =$	$(-1,1,1)$	$= 1.250\ 000\ 00$	$\bar{e} = (-2,4,0) = e + p =$	$1.265\ 625$
$z_2 =$	$(3,-2,-2)$	$= 1.280\ 000\ 00$	$\bar{y}_3 = (-9,16,0) = y_3 + 5p =$	$1.282\ 892$
$x_3 =$	$(-2,-1,4)$	$= 1.286\ 008\ 23$	$\bar{f} = (1,-1,0) = f$	
$f =$	$(1,-1,0)$	$= 1.333\ 333\ 33$	$\bar{x}_3 = (-6,11,0) = x_3 + 4p =$	$1.351\ 524$

$y_4 =$	$(-4,0,6)$	$= 1.339\ 591\ 91$	$\bar{g}es = (4,-6,0) = ges - 2p =$	$1.404\ 664$
$fis =$	$(-1,0,2)$	$= 1.388\ 888\ 89$	$\bar{f}is = (-3,6,0) = fis + 2p =$	$1.423\ 828$
$ges =$	$(2,0,-2)$	$= 1.440\ 000\ 00$	$\bar{y}_4 = (-10,18,0) = y_4 + 6p =$	$1.443\ 157$
$x_4 =$	$(-3,1,4)$	$= 1.446\ 759\ 26$	$\bar{g} = (0,1,0) = g$	
$g =$	$(0,1,0)$	$= 1.500\ 000\ 00$	$\bar{x}_4 = (-7,13,0) = x_4 + 4p =$	$1.520\ 465$
$y_5 =$	$(-5,2,6)$	$= 1.507\ 040\ 90$	$\bar{a}s = (3,-4,0) = as - p =$	$1.580\ 025$
$gis =$	$(-2,2,2)$	$= 1.562\ 500\ 00$	$\bar{g}is = (-4,8,0) = gis + 2p =$	$1.601\ 807$
$as =$	$(2,-1,-1)$	$= 1.600\ 000\ 00$	$\bar{y}_5 = (-11,20,0) = y_5 + 6p =$	$1.623\ 660$
$x_5 =$	$(-3,0,5)$	$= 1.607\ 510\ 29$	$\bar{a} = (-1,3,0) = a + p =$	$1.687\ 500$
$a =$	$(0,0,1)$	$= 1.666\ 666\ 67$	$\bar{x}_5 = (-8,15,0) = x_5 + 5p =$	$1.710\ 523$
$y_6 =$	$(-5,1,7)$	$= 1.674\ 489\ 88$	$\bar{b} = (2,-2,0) = b - p =$	$1.777\ 778$
$ais =$	$(-2,1,3)$	$= 1.736\ 111\ 11$	$\bar{a}is = (-5,10,0) = ais + 3p =$	$1.802\ 032$
$b =$	$(1,1,-1)$	$= 1.800\ 000\ 00$	$\bar{y}_6 = (-12,22,0) = y_6 + 7p =$	$1.826\ 618$
$x_6 =$	$(-4,2,5)$	$= 1.808\ 449\ 07$	$\bar{h} = (-2,5,0) = h + p =$	$1.898\ 437$
$h =$	$(-1,2,1)$	$= 1.875\ 000\ 00$	$\bar{x}_6 = (-9,17,0) = x_6 + 5p =$	$1.924\ 338$
w_1	$(-6,3,7)$	$= 1.883\ 801\ 12$	$\bar{c}^1 = (1,0,0) = c^1 =$	$2.000\ 000$
$w_2 =$	$(-3,3,3)$	$= 1.953\ 125\ 00$	$\bar{w}_2 = (-6,12,0) = w_2 + 3p =$	$2.027\ 287 > 2$
$c^1 =$	$(1,0,0)$	$= 2.000\ 000\ 00$	$\bar{w}_1 = (-13,24,0) = w_1 + 7p =$	$2.054\ 957 > 2$

Table 6.3. Relationship Between the 31-Tone Sequence in Three-Dimensional Lattice
            Space and the 29-Tone Sequence in the Two-Dimensional Pythagorean
            Lattice Space

The sequence of the 31-tone is from left to right, as is indicated by the arrows. The *images* in the Pythagorean plane of the tones of the 31-tone sequence are indicated by adding a bar to the tone. The tone sequence of the images is also indicated by arrows, however there is some backtracking involved. A line of dots connects to tones in the preceding line, a line of bars connects to tones in the following line.

Interval vectors:

$s_1 = (-7, 12, 0)$:

$s_2 = (10, 17, 0)$:

The 29 tone sequence:

$$| s_1\, s_2\, s_1\, s_2\, s_1\, | s_1\, s_2\, s_1\, s_2\, s_1\, | s_1\, s_2\, s_1\, s_2\, s_1\, | s_1\, s_2\, s_1\, s_2\, s_1\, | s_1\, s_2\, s_1\, s_2\, s_1\, | s_1\, s_2\, s_1\, s_2$$

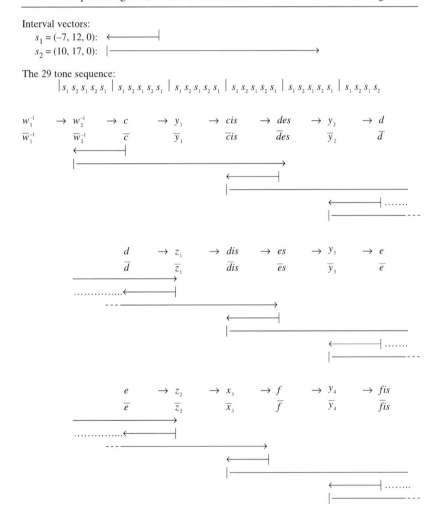

Table 6.3 (*continued*)

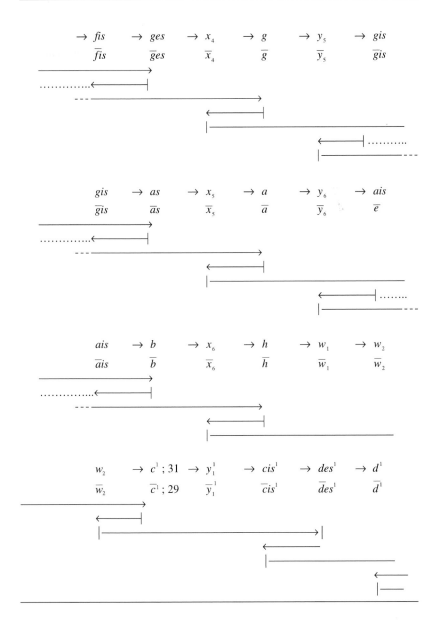

Defining the two independent vectors,

$$S_1^{-1} = s_1^{-1}p^4 \quad = (3,0,-4) = 2^3 3^4 5^{-4} \quad = 648/625 = 1.036\ 800\ 00,$$
$$S_2 = s_1 s_2 p^2 \quad = (1,1,-2) = 2^0 3^3 5^{-2} \quad = 27/25 \qquad = 1.080\ 000\ 00$$
$$(6.3a)$$

and the vector

$$S_3 = s_1^2 s_2 p^{-2} = (-2,1,2) = 2^{-3} 3^{-1} 5^2 = 25/24 \qquad = 1.041\ 666\ 67,$$
$$S_3 = S_1 S_2, \qquad s_3 = s_1 s_2, \qquad\qquad\qquad\qquad (6.3b)$$

the desired result is obtained.

The interval vectors (factors) for the 31-tone 3-dimensional music scale are obtained in terms of the constant vectors given by Eq. (6.3) as

$$\{S_1^{-1} = (3,0,-4), S_1^2 S_2 = (-5,1,6), S_1^{-1}p^{-1} = (4,-3,-3)\} \quad (6.4)$$

with

$$(S_1^{-1})^{15}(S_1^2 S_2)^{12}(S_1^{-1}p^{-1})^4 = 2, \qquad N = 31.$$

Thus, while the 2-dimensional musical system is defined by means of two interval vectors s_1 and s_2, the 3-dimensional (and thus the chromatic) musical system require three interval vectors, Eq. (6.4). These three interval vectors can be expressed in terms of the two sruti s_1, s_2, and the Pythagorean vector $p = (-1,3,-1)$ in the following way,

$$S_1^{-1} = s_1^{-1}p^4 = (3,0,-4) \qquad = 2^3 3^4 5^{-4} \qquad = 648/625$$
$$= 1.036\ 800\ 00,$$
$$S_1^2 S_2 = s_1^3 s_2 p^{-6} = (-5,1,6) \quad = 2^{-6} 3^{-5} 5^6 \qquad = 15,625/15,552$$
$$= 1.004\ 693\ 93,$$
$$S_1^{-1}p^{-1} = s_1^{-1}p^3 = (4,-3,-3) \quad = 2^7 5^{-3} \qquad = 128/125$$
$$= 1.024\ 000\ 00. \qquad\qquad\qquad\qquad (6.5)$$

Conversely, the two sruti s_1 and s_2 can be expressed in terms of S_1^{-1}, S_2 and p as

$$s_1 = S_1 p^4 \qquad\qquad = (-7,12,0) \quad = 2^{-19} 3^{12} \quad = 1.013\ 643\ 26,$$
$$s_2 = S_1^{-1} S_2 p^{-6} \qquad = (10,-17,0) \quad = 2^{27} 3^{-17} \quad = 1.039\ 318\ 25,$$
$$s_3 = s_1 s_2 = S_2 p^{-2} \quad = (3,-5,0) \qquad = 2^8 3^{-5} \qquad = 1.053\ 497\ 94.$$
$$(6.6)$$

Tables 6.1 to 6.3 list the results obtained for the 31-tone 3-dimensional tonal system.

7. Subsystems and Subscales of the Three-Dimensional 31-Tone System

In this section tonal subsystems/scales of the 31-tone system are discussed. These subsystems are obtained, like it for the Pythagorean case, by combining smaller basic intervals to form new, larger intervals, in such a manner that the sequence of new intervals exhibits regularity and closes, i.e., forms a "cycle".

The set of four combined intervals (factors)

$$\{S_3, S_1^{-1}, (S_1^{-1}p^{-1}), (S_2 p^{-1})\} \tag{7.1}$$

gives rise to the 17-tone subsystem which consists of the combined tones of the 7-tone natural diatonic musical scale, together with the 5 sharps and the 5 flats. The set of three intervals

$$\{S_3, S_2, (S_2 p^{-1})\} \tag{7.2}$$

yields both the chromatic major and the chromatic minor musical scales. Which of the two scales is obtained depends upon the order of the sequence of these intervals. The set of three intervals

$$\{(S_1 S_2^2), (S_1 S_2^2 p^{-1}), (S_2 p^{-1})\} \tag{7.3}$$

yields the natural diatonic musical scale. The set of two intervals

$$\{(S_1^2 S_2^5 p^{-2}), (S_1 S_2^2)\} \tag{7.4}$$

yields the tonal system of the ancient Greek Lyre. Still other tonal subsystems are contained as subsets of tones of the scales discussed above.

The properties of the various interval units of the subscales are

$$
\begin{aligned}
S_3 &= (-2, 1, 2) & &= 25/24, \\
S_2 &= (1, 1, -2) & &= 27/25, \\
(S_1^{-1}p^{-1}) &= (4, -3, -3) & &= 128/125, \\
(S_2 p^{-1}) &= (2, -2, -1) & &= 16/15, \\
(S_1 S_2^2) &= (-1, 2, 0) & &= 9/8, \\
(S_1 S_2^2 p^{-1}) &= (0, -1, 1) & &= 10/9, \\
(S_1^2 S_2^5 p^{-2}) &= (1, -1, 0) & &= 4/3.
\end{aligned}
\tag{7.5}
$$

Table 7.1. Summary of the Three-Dimensional 31-Tone System, Its Subsystems and Subscales

$(n, m, r) = (2/1)^n (3/2)^m (5/3)^r$, the frequency ratio of a tone,

$S_1 = (-3, 0, 4), S_1^{-1} = (3, 0, -4), S_2 = (1, 1, -2), S_3 = (-2, 1, 2) = S_1 S_2, p = (-1, 3, -1),$

$S_1^{-1} p^{-1} = (4, -3, -3), S_2 p^{-1} = (2, -2, -1), S_1^2 S_2 = (-5, 1, 6),$

$S_1^{-1} = s_1^{-1} p^4, S_2 = s_1 s_2 p^2, S_3 = S_1 S_2 = s_1^2 s_2 p^{-2},$

$s_1 = (-7, 12), s_2 = (10, -17), s_3 = s_1 s_2 = (3, -5)$

Column #1 lists the name of the tone. Column #2 defines the tone as a lattice point/vector. Column #3 defines the tone with respect to the basis $\{S_1^{-1}, S_1^2 S_2, S_2 S_1^{-1} p^{-1}\}$. Column #4 lists the interval factors/vectors between two successive tones. Column #5 lists the tones of the 17-tone system of the combined major and minor chromatic scales with basis $\{S_1 S_2, S_1^{-1}, S_2 p^{-1}, S_1^{-1} p^{-1}\}$. Column #6 lists the interval factors/vectors for column #5. Column #7 lists the 12 tones of the chromatic major scale with respect to the basis $\{S_3, S_2, S_2 p^{-1}\}$. Column #8 lists the interval factors/vectors for column #7. Column #9 lists the 12 tones of the chromatic minor with respect to the basis $\{S_2, S_3, S_2 p^{-1}\}$. Column #10 lists the interval factors/vectors for column 9.

	#1	#2	#3	#4	#5	#6	#7	#8	#9	#10
#1	c	(0,0,0)	1	S_1^{-1}	1	$S_1 S_2$	1	S_3	1	S_2
#2	y_1	(3,0,-4)	S_1^{-1}	$S_1^2 S_2$		S_1^{-1}				
#3	cis	(-2,1,2)	$S_1 S_2$	S_1^{-1}	$S_1 S_2$	S_1^{-1}	S_3	S_2		
#4	des	(1,1,-2)	S_2	$S_1^2 S_2$	S_2	$S_1 S_2$			S_2	S_3
#5	y_2	(-4,2,4)	$S_1^2 S_2^2$	S_1^{-1}						
#6	d	(-1,2,0)	$S_1 S_2^2$	S_1^{-1}	$S_1 S_2^2$	$S_1 S_2$	$S_2 S_3$	S_3	$S_2 S_3$	$S_2 p^{-1}$
#7	z_1	(2,2,-4)	S_2^2	$S_1^2 S_2$						
#8	dis	(-3,3,2)	$S_1^2 S_2^3$	$S_1^{-1} p^{-1}$	$S_1^2 S_2^3$	$S_1^{-1} p^{-1}$	$S_2 S_3^2$	$S_2 p^{-1}$		
#9	es	(1,0,-1)	$S_1 S_2^3 p^{-1}$	$S_1^2 S_2$	$S_1 S_2^3 p^{-1}$	$S_1 S_2$			$S_2^2 S_3 p^{-1}$	S_3
#10	y_3	(-4,1,5)	$S_1^3 S_2^4 p^{-1}$	S_1^{-1}	S_1^{-1}					

#		(a,b,c)	$S_1^2 S_2^4 p^{-1}$	$S_1^{-1} p^{-1}$	$S_1^2 S_2^4 p^{-1}$	$S_2 p^{-1}$	$S_2^2 S_3^2 p^{-1}$	$S_2 p^{-1}$	$S_2^2 S_3^2 p^{-1}$	$S_2 p^{-1}$
#11	e	$(-1,1,1)$	$S_1^2 S_2^4 p^{-1}$	$S_1^{-1} p^{-1}$	$S_1^2 S_2^4 p^{-1}$	$S_2 p^{-1}$	$S_2^2 S_3^2 p^{-1}$	$S_2 p^{-1}$	$S_2^2 S_3^2 p^{-1}$	$S_2 p^{-1}$
#12	z_2	$(3,-2,-2)$	$S_1^1 S_2^4 p^{-2}$	$S_1^2 S_2$						S_2
#13	x_3	$(-2,-1,4)$	$S_1^3 S_2^5 p^{-2}$	S_1^{-1}	$S_1^3 S_2^5 p^{-2}$	$S_1 S_2$	$S_2^3 S_3^2 p^{-2}$	S_3	$S_2^3 S_3^2 p^{-2}$	S_3
#14	f	$(1,-1,0)$	$S_1^2 S_2^5 p^{-2}$	$S_1^2 S_2$	$S_1^2 S_2^5 p^{-2}$	S_1^{-1}		S_2		
#15	y_4	$(-4,0,6)$	$S_1^4 S_2^6 p^{-2}$	S_1^{-1}	$S_1^4 S_2^6 p^{-2}$	$S_1 S_2$				
#16	fis	$(-1,0,2)$	$S_1^3 S_2^6 p^{-2}$	S_1^{-1}	$S_1^3 S_2^6 p^{-2}$	S_1^{-1}	$S_2^4 S_3^2 p^{-2}$	S_3	$S_2^3 S_3^3 p^{-2}$	S_3
#17	ges	$(2,0,-2)$	$S_1^2 S_2^6 p^{-2}$	$S_1^2 S_2$	$S_1^2 S_2^6 p^{-2}$	$S_1 S_2$		S_2		S_2
#18	x_4	$(-3,1,4)$	$S_1^4 S_2^7 p^{-2}$	S_1^{-1}	$S_1^3 S_2^7 p^{-2}$	S_1^{-1}	$S_2^4 S_3^3 p^{-2}$	$S_2 p^{-1}$	$S_2^4 S_3^3 p^{-2}$	$S_2 p^{-1}$
#19	g	$(0,1,0)$	$S_1^3 S_2^7 p^{-2}$	$S_1^2 S_2$	$S_1^3 S_2^7 p^{-2}$	$S_1 S_2$				
#20	y_5	$(-5,2,6)$	$S_1^5 S_2^8 p^{-2}$	$S_1^{-1} p^{-1}$	$S_1^4 S_2^8 p^{-2}$	$S_1^{-1} p^{-1}$	$S_2^5 S_3^3 p^{-2}$		$S_2^5 S_3^3 p^{-3}$	
#21	gis	$(-2,2,2)$	$S_1^4 S_2^8 p^{-2}$	S_1^{-1}	$S_1^3 S_2^8 p^{-2}$	S_1^{-1}				
#22	as	$(2,-1,-1)$	$S_1^3 S_2^8 p^{-3}$	$S_1^2 S_2$	$S_1^3 S_2^8 p^{-3}$	$S_1 S_2$		S_2		S_3
#23	x_5	$(-3,0,5)$	$S_1^5 S_2^9 p^{-3}$	S_1^{-1}	$S_1^5 S_2^9 p^{-3}$	S_1^{-1}	$S_2^5 S_3^4 p^{-3}$		$S_2^5 S_3^4 p^{-3}$	
#24	a	$(0,0,1)$	$S_1^4 S_2^9 p^{-3}$	$S_1^2 S_2$	$S_1^4 S_2^9 p^{-3}$	$S_1 S_2$	$S_2^5 S_3^4 p^{-3}$	$S_1 S_2$	$S_2^5 S_3^4 p^{-3}$	S_2
#25	y_6	$(-5,1,7)$	$S_1^6 S_2^{10} p^{-3}$	S_1^{-1}	$S_1^5 S_2^{10} p^{-3}$	S_1^{-1}	$S_2^6 S_3^4 p^{-3}$		$S_2^5 S_3^5 p^{-3}$	
#26	ais	$(-2,1,3)$	$S_1^5 S_2^{10} p^{-3}$	S_1^{-1}	$S_1^5 S_2^{10} p^{-3}$	S_1^{-1}				
#27	b	$(1,1,-1)$	$S_1^4 S_2^{10} p^{-3}$	$S_1^2 S_2$	$S_1^4 S_2^{10} p^{-3}$	$S_1 S_2$		$S_1 S_2$		S_3
#28	x_6	$(-4,2,5)$	$S_1^6 S_2^{11} p^{-3}$	S_1^{-1}	$S_1^6 S_2^{11} p^{-3}$	S_1^{-1}	$S_2^6 S_3^5 p^{-3}$		$S_2^6 S_3^5 p^{-3}$	
#29	h	$(-1,2,1)$	$S_1^5 S_2^{11} p^{-3}$	$S_1^2 S_2$	$S_1^5 S_2^{11} p^{-3}$	$S_1^2 S_2$	$S_2^6 S_3^5 p^{-3}$	$S_2 p^{-1}$	$S_2^6 S_3^5 p^{-3}$	$S_2 p^{-1}$
#30	w_1	$(-6,3,7)$	$S_1^7 S_2^{12} p^{-3}$	S_1^{-1}	$S_1^7 S_2^{12} p^{-3}$	S_1^{-1}				
#31	w_2	$(-3,3,3)$	$S_1^6 S_2^{12} p^{-3}$	$S_1^{-1} p^{-1}$	$S_1^6 S_2^{12} p^{-3}$	$S_1^{-1} p^{-1}$	$S_2^7 S_3^5 p^{-4} = 2$		$S_2^7 S_3^5 p^{-4} = 2$	
#32	c^1	$(1,0,0)$	$S_1^5 S_2^{12} p^{-4} = 2$		$S_1^5 S_2^{12} p^{-4} = 2$					

(continued)

Table 7.1 (*continued*)

Column #13 lists the 7 tones of the natural diatonic scale with respect to the basis $\{T_1 = S_1 S_2^2, T_2 = S_1 S_2 p^{-1}, S = S_2 p^{-1}\}$. Column #14 lists the interval factors/vectors for column #12. Column #15 lists the 7 tones of the natural diatonic scale with respect to the basis $\{s_1^3 s_2^2, s_1^3 s_2^2 p^{-1}, s_1 s_2 p\}$. Column #16 lists the interval factors/vectors for column #14. Columns #17 and #18 list the tones and interval factors/vectors for the 3 tones of the ancient Greek Lyre

Multiplication/division for the numerical values (frequency ratios) corresponds to addition/subtraction of the vector components (the exponents n, m, r). Thus, an expression like $S_2^2 S_2$ can be read as an ordinary product of numbers (frequency ratios), or as a sum of vectors $(-3, 0, 4) + (-3, 0, 4) + (1, 1, -2) = (-5, 1, 6)$

	#11	#12	#13	#14	#15	#16	#17	#18
#1	1	$(0, 0, 0)$	1	$T_1 = S_1 S_2^2$	1	$T_1 = s_1^3 s_2^2$	1	$T_1 T_2 S = s_1^7 s_2^5$
#2	y_1	$(3, 0, -4)$						
#3	cis	$(-2, 1, 2)$						
#4	des	$(1, 1, -2)$						
#5	y_2	$(-4, 2, 4)$						
#6	d	$(-1, 2, 0)$	$S_1 S_2^2$	$T_2 = S_1 S_2^2 p^{-1}$	$s_1^3 s_2^2$	$T_2 = s_1^3 s_2^2 p^{-1}$		
#7	z_1	$(2, 2, -4)$						
#8	dis	$(-3, 3, 2)$						
#9	es	$(1, 0, -1)$						
#10	y_3	$(-4, 1, 5)$						

#								
#11	e	$(-1,1,1)$	$S_1^2 S_2^4 p^{-1}$	$S = S_2 p^{-1}$	$s_1^6 s_2^4 p^{-1}$	$S = s_1 s_2 p$		
#12	z_2	$(3,-2,-2)$						
#13	x_3	$(-2,-1,4)$						
#14	f	$(1,-1,0)$	$S_1^2 S_2^5 p^{-2}$	$T_1 = S_1 S_2^2$	$s_1^7 s_2^5$	$T_1 = s_1^3 s_2^2$	$s_1^7 s_2^5$	$T_1 = s_1^3 s_2^2$
#15	y_4	$(-4,0,6)$						
#16	fis	$(-1,0,2)$						
#17	ges	$(2,0,-2)$						
#18	x_4	$(-3,1,4)$						
#19	g	$(0,1,0)$	$S_1^3 S_2^7 p^{-2}$	$T_2 = S_1 S_2^2 p^{-1}$	$s_1^{10} s_2^7$	$T_2 = s_1^3 s_2^2 p^{-1}$	$s_1^{10} s_2^7$	$T_1 T_2 S = s_1^7 s_2^5$
#20	y_5	$(-5,2,6)$						
#21	gis	$(-2,2,2)$						
#22	as	$(2,-1,-1)$						
#23	x_5	$(-3,0,5)$						
#24	a	$(0,0,1)$	$S_1^4 S_2^9 p^{-3}$	$T_1 = S_1 S_2^2$	$s_1^{13} s_2^9 p^{-1}$	$T_1 = s_1^3 s_2^2$		
#25	y_6	$(-5,1,7)$						
#26	ais	$(-2,1,3)$						
#27	b	$(1,1,-1)$						
#28	x_6	$(-4,2,5)$						
#29	h	$(-1,2,1)$	$S_1^5 S_2^{11} p^{-3}$	$S = S_2 p^{-1}$	$s_1^{16} s_2^{11} p^{-1}$	$S = s_1 s_2 p$		
#30	w_1	$(-6,3,7)$						
#31	w_2	$(-3,3,3)$						
#32	c^1	$(1,0,0)$	$S_1^5 S_2^{12} p^{-4} = 2$		$s_1^{17} s_2^{12} = 2$		$s_1^{17} s_2^{12} = 2$	$(T_1 T_2 S)^2 T_1 = 2$

Expressed in terms of the two sruti s_1, s_2, and the Pythagorean vector p, these intervals are given as

$$S_3 = (s_1^2 s_2 p^{-2}),$$
$$S_2 = (s_1 s_2 p^2),$$
$$(S_1^{-1} p^{-1}) = (s_1^{-1} p^3),$$
$$(S_2 p^{-1}) = (s_1 s_2 p) = S,$$
$$(S_1 S_2^2) = (s_1^3 s_2^2) = t = T_1,$$
$$(S_1 S_2^2 p^{-1}) = (s_1^3 s_2^2 p^{-1}) = T_2,$$
$$(S_1^2 S_2^5 p^{-2}) = (s_1^7 s_2^5),$$
$$p = T_1/T_2 = S/s_1 s_2, \qquad p^2 = S_2/s_1 s_2, \qquad s = s_3. \tag{7.6}$$

For the 17-tone sub-set of the chromatic musical scales holds

$$S_3^{10}(S_1^{-1})^3 (S_1^{-1} p^{-1})^2 (S_2 p^{-1})^2 = (1, 0, 0). \tag{7.7}$$

For the chromatic major and minor musical scales holds

$$S_3^5 S_2^3 (S_2 p^{-1})^4 = (1, 0, 0). \tag{7.8}$$

For the diatonic musical scale, with the two tones T_1 and T_2, and the semitone S, holds

$$\begin{aligned}
T_1 &= (S_1 S_2^2) &&= (s_1^3 s_2^2) &&= (-1, 2, 0), \\
T_2 &= (S_1 S_2^2 p^{-1}) &&= (s_1^3 s_2^2 p^{-1}) &&= (0, -1, 1), \\
S &= (S_2 p^{-1}) &&= (s_1 s_2 p) &&= (2, -2, -1),
\end{aligned} \tag{7.9}$$

with

$$T_1^3 T_2^2 S^2 = (1, 0, 0) = (2/1)^1 (3/2)^0 (5/3)^0 = 2 \tag{7.10}$$

and

$$p = T_1 T_2^{-1} = (S_2/(s_1 s_2))^{1/2} = (-1, 3, -1) = 81/80. \tag{7.11}$$

The last equation expresses the fact that the Pythagorean vector p is the key to the relationship between the intervals of the Carnatic and the natural diatonic musical scales. Note that the factor p^{-2} cancels in the expression $(S_1^2 S_2^5 p^{-2})$, Eq. (7.5). This is the reason for the fact that the scale of the ancient Greek Lyre is contained in both the chromatic and the Carnatic musical systems. Table 7.1 contains a summary of these results.

It can also be verified in Table 7.1 that the tone sequence of column #9 maps, along the vector p, upon the standard 12-tone sequence in the Pythagorean plane

$$(0,0), (3,-5), (-1,2), (2,-3), (-2,4), (1,-1), (4,-6), (0,1),$$
$$(3,-4), (-1,3), (2,-2), (-2,5),$$

while the sequence of tones given by the column #7 maps upon the 12-tone 2-dimensional sequence

$$(0,0), (-4,7), (-1,2), (-5,9), (-2,4), (1,-1), (-3,6), (0,1),$$
$$(-4,8), (-1,3), (-5,10), (-2,5).$$

8. Correlation of Tone Sequences in Two-Dimensional Pythagorean Plane with the Three-Dimensional 31-Tone Sequence

The images of the chromatic major and minor 3-dimensional scales in the Pythagorean plane will be called "Pythagorean major" and "Pythagorean minor" musical scales. The tones of the chromatic major and minor 3-dimensional scales are, by means of the map along the Pythagorean vector $p = (-1, 3, -1) = 81/80$, uniquely correlated with the tones of the 2-dimensional "Pythagorean major" and "Pythagorean minor" scales.

As was pointed out in Sect. 1, the set of 3-dimensional lattice points

$$(n + r, m - 3r, r), n, m = \text{fixed}, \qquad r = 0, \pm 1, \pm 2, \pm 3, \ldots \quad (8.1)$$

is mapped onto the 2-dimensional Pythagorean lattice point

$$(n, m, 0) = (n + r, m - 3r, r) + r(-1, 3, -1),$$

$$r = 0, \pm 1, \pm 2, \pm 3, \ldots. \quad (8.2)$$

This map is somewhat analogous to the map of the (2/1)-octave tones c^n, $n = 0, \pm 1, \pm 2, \pm 3, \ldots$, along the vector $(1, 0, 0)$, upon the reference tone $\nu_0 = c_0 = c^0 = (0, 0, 0) = 1$, namely

$$c^n/(2^n c_0) = 1, \qquad n = 0, \pm 1, \pm 2, \pm 3, \ldots \quad (8.3a)$$

or equivalently

$$(0, 0, 0) = (n, 0, 0) - n(1, 0, 0), \qquad n = 0, \pm 1, \pm 2, \pm 3, \ldots. \quad (8.3b)$$

That is, Eq. (8.3) represents a map of the (2/1)-octave tones c^n, which can be considered to be musical tones equivalent to the reference tone c_0 (except for pitch, i.e. a scaling factor which cancels in musical

Table 8.1. The Tone Sequence c, a, d^1, g, c in the Three-Dimensional Tone Lattice and Its Projection into the Two-Dimensional Pythagorean Plane

Column #1: names of sequence of 3-dimensional lattice tones. Column #2: the lattice tones. Column #3: the frequency ratio of the lattice points with respect to the tone c. Column #4: the interval vectors between two successive lattice tones. Column #5: the numerical value (frequency ratio) of the interval vectors. Column #6: the relationship between the 3-dimensional lattice tones to their image in the 2-dimensional Pythagorean plane. Column #7: the 2-dimensional images in the Pythagorean plane. Column #8: the images of the 3-dimensional interval vectors in the Pythagorean plane. Column #9: the frequency ratios for the interval vectors of column #8. Columns #7 and #8 show that the images form a closed sequence of tones.

#1	#2	#3	#4	#5	#6	#7	#8	#9
c/c	$(0,0,0)$	1	$(0,0,1)$	5/3	$c = (0,0,0)$	$(0,0)$	$(-1,3)$	27/16
a/c	$(0,0,1)$	5/3	$(1,-1,0)$	4/3	$a = (-1,3,0) - p = (27/16)(80/81) = 5/3$	$(-1,3)$	$(1,-1)$	4/3
x/c	$(1,-1,1)$	20/9	$(0,-1,0)$	2/3	$x = (0,2,0) - p = (9/4)(80/81) = 20/9$	$(0,2)$	$(0,-1)$	2/3
y/c	$(1,-2,1)$	40/27	$(0,-1,0)$	2/3	$y = (0,1,0) - p = (3/2)(80/81) = 40/27$	$(0,1)$	$(0,-1)$	2/3
z/c	$(1,-3,1)$	80/81			$z = (0,0,0) - p = (1/1)(80/81)$	$(0,0)$		

frequency ratios), upon the basic octave tone $c_0 = c$. Thus the lattice points given by Eq. (8.1) will also be called equivalent lattice points (with respect to translations by the Pythagorean vector p).

The members of an equivalent set of lattice points are then distinguished from each other by their location on distinct Pythagorean planes labeled by r,

$$(n, m, r), n, m = 0, \pm 1, \pm 2, \pm 3, \ldots. \tag{8.4}$$

The $r = 0$ Pythagorean plane is the basic plane, while the other Pythagorean planes are equivalent parallel planes. These parallel planes can be considered to result from a translation, by integer multiples, by the Pythagorean vector $p = (-1, 3, -1)$. Thus the musical tones on two planes, related by a translation by an integer multiple of the vector $p = (-1, 3, -1)$, are equivalent musical tones (with respect to the translation p). In particular, the points equivalent to the origin $(0, 0, 0) = c^0/c_0 = 1$ of the basic Pythagorean lattice are given by the chromatic lattice points on the r-th Pythagorean-Carnatic plane

$$(0,0,0) + r(-1,3,-1) = r(-1,3,-1), \quad r = 0, \pm 1, \pm 2, \pm 3, \ldots.$$
$$1 \quad \times \quad (81/80)^r \ = (81/80)^r$$

$$\tag{8.5}$$

A familiar example will illustrate this point, PIERCE, ref. [12]. The frequency ratios of the tones/keys of the sequence c, a, d^1, g, c, with respect to the tone c, are

$$c/c = 1, \quad a/c = 5/3, \quad d^1/c = 9/4, \quad g/c = 3/2, \quad c/c = 1.$$

Thus, the interval factors/vectors connecting the corresponding successive 3-dimensional lattice tones are

$$c; \quad (0,0,1) \rightarrow a; \ (1,-1,0) \rightarrow x; \ (0,-1,0) \rightarrow y; \ (0,-1,0) \rightarrow z = p^1.$$
$$(0,0,0) \qquad (0,0,1) \qquad (1,-1,1) \qquad (1,-2,1) \qquad (1,-3,1)$$

The vectors to the right of the tones are the interval vectors/factors connecting neighboring tones, while the vectors under the tones represent the tones as 3-dimensional lattice tones. It is seen that the sequence of the 3-dimensional lattice tones ends up not at the tone $c = (0,0,0)$ but at the lattice tone $p^{-1} = (1, -3, 1) = 80/81$, p the syntonic comma.

The Pythagorean tones contained in Table 8.1 are

$\bar{c}$	$= (0,0,0)$	$= (0,0)$	$= 1$	$= c,$
$\bar{a}$	$= (-1,3,0)$	$= (-1,3)$	$= 27/16$	$= a + p,$
$\bar{d}^1 = d^1$	$= (0,2,0)$	$= (0,2)$	$= 9/4$	$= x + p,$
$\bar{g} = g$	$= (0,1,0)$	$= (0,1)$	$= 3/2$	$= y + p,$

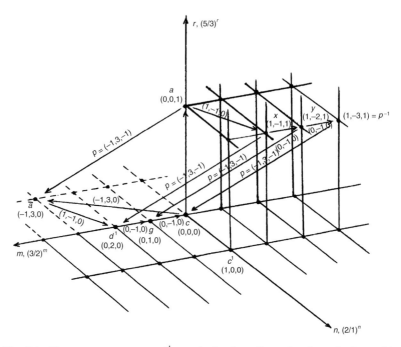

Fig. 8.1. The tone sequence c, a, d^1, g, c in the three-dimensional tone lattice and its projection into the Pythagorean plane. This figure shows a small part of the 3-dimensional tonal lattice around the tone $(0,0,0) = 1$, i.e. the tone c. The lattice point distances are *not* to scale but have been chosen such as to admit a clearer graphical representation. It is seen that the sequence of tones c, a, d^1, g, c actually corresponds to the tone sequence $(0,0,0)$, $(0,0,1)$, $(1,-1,1)$, $(1,-2,1)$, $(1,-3,1)$, and thus leads to the tone $p^{-1} = (1,-3,1) = 80/81$, and not to the original tone $(0,0,0) = 1$. The mapping, along the vector p, of the 31-tone 3-dimensional sequence c, a, d^1, g, c from the chromatic lattice into the $r = 0$ Pythagorean plane, yields the image sequence ("the shadow cast in the plane by the 3-dimensional tones") $\bar{c} = c, \bar{a} = a, \bar{d}^1 = d^1, \bar{g} = g, \bar{c} = c$ which *does* close. It can also be seen that the numerical values associated with these tones can be mapped onto the $(2/1)$-octave line $(0, \infty)$, the standard representation for the musical tones. These *numerical values* of the musical tones, however, do not fully represent the musical tones, but represent merely one property of the musical tones which are given as *lattice points* (n, m, r)

with

$$p = (-1, 3, -1).$$

The tones x, y, z are defined in the table (the reference tone c is set equal to 1, except where a special point needs to be made). A graphical representation of Table 8.1 is given in Fig. 8.1.

Thus, the tone c which is reached by this sequence of lattice tones is seen to be a tone c which lies in the $r = 1$ Pythagorean plane, a tone which is equivalent, with respect to the Pythagorean vector p, to the tone c in the basic Pythagorean plane, $r = 0$. The cycle of tones does not close since the tone $a = 5/3$ leads to a tone outside the basic Pythagorean-Carnatic plane. The sequence of images $\bar{c} = c$, $\bar{a}$, $\bar{d}^1$, $\bar{g} = g$, $\bar{c} = c$, however, closes in the Pythagorean plane. Thus the "inner structure" of the musical tones (n, m, r), which characterizes the musical tones in terms of a "three octave system", the standard $(2/1)$-octave system and two rescaled octave systems, the $(3/2)$-octave system and the $(5/3)$-octave system, provides a deeper insight into musical properties of tones, and tone systems, than their associated numerical values alone (the frequency ratios) can provide. See Fig. 8.1. (See also Figs. 8, 12 and 15 of [1]).

The correspondence between the chromatic major and minor tone scales and the "Carnatic major" and "Carnatic minor" tone scale is given by

$$
\begin{array}{llll}
\bar{c} & = (0,0,0) & = c & \\
\bar{cis} & = (-4,7,0) & = cis + 2p & = (-2,1,2) + 2p \\
\bar{des} & = (3,-5,0) & = des - 2p & = (1,1,-2) - 2p \\
\bar{d} & = (-1,2,0) & = d & \\
\bar{es} & = (2,-3,0) & = es - p & = (1,0,-1) - p \\
\bar{dis} & = (5,-9,0) & = dis + 2p & = (-3,3,2) + 2p \\
\bar{e} & = (-2,4,0) & = e + p & = (-1,1,1) + p \\
\bar{f} & = (1,-1,0) & = f & \\
\bar{ges} & = (4,-6,0) & = ges - 2p & = (2,0,-2) - 2p \\
\bar{fis} & = (-3,6,0) & = fis + 2p & = (-1,0,2) + 2p \\
\bar{g} & = (0,1,0) & = g & \\
\bar{as} & = (3,-4,0) & = as - p & = (2,-1,-1) - p \\
\bar{gis} & = (-4,8,0) & = gis + 2p & = (-2,2,2) + 2p \\
\bar{a} & = (-1,3,0) & = a + p & = (0,0,1) + p \\
\bar{b} & = (2,-2,0) & = b - p & = (1,1,-1) - p \\
\bar{ais} & = (-5,10,0) & = ais + 3p & = (-2,1,3) + 3p \\
\bar{h} & = (-2,5,0) & = h + p & = (-1,2,1) + p \\
\hline
\bar{c}^1 & = (1,0,0) & = c^1 & \hfill (8.6)
\end{array}
$$

The list above shows that the frequencies of the pure tones of the chromatic major and the chromatic minor musical scales are, in the corresponding tones of the Carnatic system, modified by factors p^r,

$$
p^r = (81/80)^r, \qquad r = 0, \pm 1, \pm 2, \pm 3, \dots, \qquad (8.7)
$$

where r indicates the r-th level of the Pythagorean plane, with $r = 0$ denoting the basic Pythagorean plane. That is, the r-th Pythagorean plane is equivalent to the basic Pythagorean-Carnatic plane, modulo the vector p, and is obtained from it by a translation (a shift) by the vector

$$p^n = n(-1, 3, -1) = (-n, 3n, -n). \qquad (8.8)$$

For example, the chromatic tone $cis = 25/24$, which corresponds to the lattice point $= (-2, 1, 2)$ on the $r = 2$ Pythagorean plane, is translated by the vector $2p = 2(-1, 3, -1) = (81/80)^2$ into the $r = 0$ basic Pythagorean lattice point $\bar{c}is = (-4, 7, 0)$,

$$\bar{c}is = (2/1)^{-4}(3/2)^7 = ((2/1)^{-2}(3/2)^1(5/3)^2)((2/1)^{-2}(3/2)^6(5/3)^{-2})$$
$$= cis\, p^2. \qquad (8.9)$$

The chromatic interval factors for the chromatic major and minor scales are

$$\{(S_1 S_2), S_2, (S_2 p^{-1})\}, \qquad S_1 = (-3, 0, 4), \qquad S_2 = (1, 1, -2),$$

$$p = (-1, 3, -1). \qquad (8.10)$$

These *three* interval factors are projected into the basic Pythagorean plane,

$$(S_1 S_2)p^2 = (s_1^2 s_2), \qquad S_2 p^{-2} = (s_1 s_2), \qquad (S_2 p^{-1})p^{-1} = (s_1 s_2)$$

to become the *two* interval factors for the "Carnatic major" and "Carnatic minor" scales

$$\{(s_1^2 s_2), (s_1 s_2)\}. \qquad (8.11)$$

Table 8.2. The Images of the Three-Dimensional Major and Minor Chromatic Musical Scales in the Two-Dimensional Pythagorean Plane

$s_1 = (-7, 12, 0)$,	$s_2 = (10, -17, 0)$,		$s = s_3 = s_1 s_2$,
$\bar{c} = c;\ s_3 \rightarrow \bar{d}es;\ s_1$	$\rightarrow \bar{c}is;\ s_3$	$\rightarrow \bar{d} = d$	$;t = (-1, 2, 0) = 9/8$
$\bar{d};\ s_3 \quad \rightarrow \bar{e}s;\ s_1$	$\rightarrow \bar{d}is;\ s_3$	$\rightarrow \bar{e}$	$;t = s_1 s_3^2 = s_3^3 s_2^2$
$\bar{e};\ s_3 \quad \rightarrow \bar{f} = f;\ s_1$	$\rightarrow \bar{g}es;\ s_3$	$\rightarrow \bar{f}is$	$;t$
$\bar{f}is;\ s_3 \quad \rightarrow \bar{g} = g;\ s_1$	$\rightarrow \bar{a}s;\ s_3$	$\rightarrow \bar{g}is$	$;t$
$\bar{g}is;\ s_3 \quad \rightarrow \bar{a};\ s_1$	$\rightarrow \bar{b};\ s_3$	$\rightarrow \bar{a}is$	$;t$
$\bar{a}is;\ s_3 \quad \rightarrow \bar{h};\ s_3$	$\rightarrow \bar{c}^1 = c^1;$		$;s^2$
$s_1^5 s_3^{12} = 2,$	$5(-7, 12, 0) + 12(3, -5, 0) = (1, 0, 0)$		
$(= t^5 s^2)$	$(= 5(-1, 2, 0) + 2(3, -5, 0))$		

Vice versa, the *three* intervals

$$\{(s_1^2 s_2)p^{-2} = (S_1 S_2), (s_1 s_2)p^2, (s_1 s_2)p\} \qquad (8.12)$$

represent an embedding of the *two* Carnatic intervals in 3-dimensional lattice space.

The images of the chromatic minor and the major musical scales are given in Table 8.1. Tables 8.2–8.4 illustrate the correlation of the Carnatic tone sequence with respect to the 31-tone 3-dimensional tone sequence.

The images of the chromatic major and minor musical scales in the Pythagorean plane are given by the ordered sequences:

$$|(s_3 s_1)s_3|(s_3 s_1)s_3|s_3(s_3 s_1)|s_3(s_3 s_1)|s_3(s_3 s_1)|s_3 s_3$$

for the "Carnatic major" scale/system, and

$$|s_3(s_3 s_1)|s_3(s_3 s_1)|s_3 s_3(s_1|s_3)s_3(s_1|s_3)s_3(s_1|s_3)s_3,$$

$$\text{containing the tone } \bar{g}es,$$

$$|s_3(s_3 s_1)|s_3(s_3 s_1)|s_3(s_3 s_1)|(s_3 s_1)s_3|(s_3 s_1)s_3|(s_3 s_1)s_3,$$

$$\text{containing the tone } \bar{f}is,$$

for the "Carnatic minor" scale/system. See also Table 10.1.

9. Linearization – the Unit Interval and the Cent

In [1] two relationships for sound frequencies were discussed. The frequency ratios ν/ν_0 within the basic, $n = 0$, $(2/1)$-octave interval (as well as any other $(2/1)$-octave interval) can be expressed in the two forms

$$\nu/\nu_0 = 1 + \delta/2\pi, \qquad 0 \le \delta/2\pi \le 1 \qquad (9.1)$$

and in the form

$$\nu/\nu_0 = 2^{\xi/2\pi}, \qquad 0 \le \xi/2\pi \le 1, \qquad (9.2)$$

with ν_0 an arbitrarily chosen, but fixed, reference frequency. Both equations cover the frequency values ν/ν_0 of the closed interval $[1, 2]$, but obviously in a linear and in an exponential manner, respectively. Thus for a fixed parameter value α,

$$0 < \delta/2\pi = \xi/2\pi = \alpha < 1, \qquad (9.3)$$

it holds

$$\nu/\nu_0 = 1 + \alpha \ne \nu'/\nu_0 = 2^\alpha, \qquad (9.4)$$

Table 8.3. Relationship Between the 31-Tone Three-Dimensional Sequence and the
        23/22-Tone Carnatic Scales/Systems in the Pythagorean Plane

The sequence of the micro-chromatic tones is from left to right, as is indicated by the
arrows. The 2-dimensional Pythagorean images of the 31-tone, 3-dimensional
system, are characterized by bars. The sequence of these tones is also indicated by
arrows, however there is some backtracking involved.

Interval vectors:

$S_3 = (3, -5, 0)$:

$S_1 = (-7, 12, 0)$:

$S_2 = (10, -17, 0)$:

| w_1^{-1} | $\to$ | w_2^{-1} | $\to$ | c | $\to$ | y_1 | $\to$ | cis | $\to$ | des | $\to$ | y_2 | $\to$ | d |
| $\overline{w_1}^{-1}$ | | $\overline{w_2}^{-1}$ | | $\overline{c}$ | | $\overline{y_1}$ | | $\overline{cis}$ | | $\overline{des}$ | | $\overline{y_2}$ | | $\overline{d}$ |

| d | $\to$ | z_1 | $\to$ | dis | $\to$ | es | $\to$ | y_3 | $\to$ | e |
| $\overline{d}$ | | $\overline{z_1}$ | | $\overline{dis}$ | | $\overline{es}$ | | $\overline{y_3}$ | | $\overline{e}$ |

| e | $\to$ | z_2 | $\to$ | x_3 | $\to$ | f | $\to$ | y_4 | $\to$ | fis |
| $\overline{e}$ | | $\overline{z_2}$ | | $\overline{x_3}$ | | $\overline{f}$ | | $\overline{y_4}$ | | $\overline{fis}$ |

| fis | $\to$ | ges | $\to$ | x_4 | $\to$ | g | $\to$ | y_5 | $\to$ | gis |
| $\overline{fis}$ | | $\overline{ges}$ | | $\overline{x_4}$ | | $\overline{g}$ | | $\overline{y_5}$ | | $\overline{gis}$ |

Table 8.3 (*continued*)

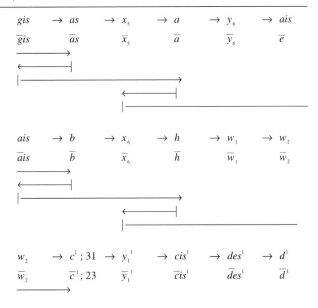

for the *23-tone Carnatic scale/tonal system*, and

for the *22-tone Carnatic scale/tonal system*.

i.e. the frequency ratios ν/ν_0 and ν'/ν_0 are not equal. They are equal only at the two end points of the interval, namely for

$$\alpha = \delta/2\pi = \xi/2\pi = 0 \text{ and } 1,$$

with

$$\nu/\nu_0 = \nu'/\nu_0 = 1 \text{ and } 2.$$

Table 8.4. Relationship Between the 31-Tone Three-Dimensional Tonal System and the Set of 17 Tones of the Two-Dimensional "Carnatic Major" and "Carnatic Minor" Scales

The sequence of the 3-dimensional tones is from left to right, as is indicated by the arrows. The Pythagorean 2-dimensional images of the 3-dimensional tones are characterized by a bar. The sequence of the 2-dimensional tones is also indicated by arrows, however there is some backtracking involved.

w_1^{-1} → w_2^{-1} → c → y_1 → cis → des → y_2 → d

$\overline{w}_1^{-1}$ $\overline{w}_2^{-1}$ $\overline{c}$ $\overline{y}_1$ $\overline{cis}$ $\overline{des}$ $\overline{y}_2$ $\overline{d}$

$s_3 = (3, -5, 0) \,|$ ———————————————————→

← ——————— $|\ s_1 = (-7, 12, 0)$

$s_3 = (3, -5, 0) \,|$ ———————————————————→

d → z_1 → dis → es → y_3 → e

$\overline{d}$ $\overline{z}_1$ $\overline{dis}$ $\overline{es}$ $\overline{y}_3$ $\overline{e}$

$|$ ——————————————→

← ——————— $|$

$|$ ——————————————————→

e → z_2 → x_3 → f → y_4 → fis

$\overline{e}$ $\overline{z}_2$ $\overline{x}_3$ $\overline{f}$ $\overline{y}_4$ $\overline{fis}$

$|$ ——————————————→$|$

fis → ges → x_4 → g → y_5 → gis

$\overline{fis}$ $\overline{ges}$ $\overline{x}_4$ $\overline{g}$ $\overline{y}_5$ $\overline{gis}$

——————→

← ——————— $|$

$|$ ——————————————→$|$

gis → as → x_5 → a → y_6 → ais

$\overline{gis}$ $\overline{as}$ $\overline{x}_5$ $\overline{a}$ $\overline{y}_6$ $\overline{e}$

——————→

← ——————— $|$

$|$ ——————————————→$|$

ais → b → x_6 → h → w_1 → w_2

$\overline{ais}$ $\overline{b}$ $\overline{x}_6$ $\overline{h}$ $\overline{w}_1$ $\overline{w}_2$

——————→

← ——————— $|$

$|$ ——————————————→$|$

w_2 → c^1; 31 → y_1^1 → cis^1 → des^1 → d^1

$\overline{w}_2$ $\overline{c}^1$; 17 $\overline{y}_1^1$ $\overline{cis}^1$ $\overline{des}^1$ $\overline{d}^1$

——————→

Table 8.5. Relationship Between the 31-Tone Three-Dimensional Sequence and the 7-Tone Two-Dimensional Pythagorean Sequence of Tones

The sequence of the 3-dimensional tones is from left to right, as is indicated by the arrows. The Pythagorean 2-dimensional images of the 3-dimensional 31-tone system are characterized by a bar. $(s_1 s_3) = (-4, 7, 0)$, $s_3 = (3, -5, 0)$

$$
\begin{array}{cccccccc}
w_1^{-1} & \rightarrow w_2^{-1} & \rightarrow c & \rightarrow y_1 & \rightarrow cis & \rightarrow des & \rightarrow y_2 & \rightarrow d \\
\overline{w_1^{-1}} & \overline{w_2^{-1}} & \overline{c} & \overline{y_1} & \overline{cis} & \overline{des} & \overline{y_2} & \overline{d}
\end{array}
$$

$(s_1 s_3) \;|\!\!\longrightarrow\!\longrightarrow$

$$
\begin{array}{cccccc}
d & \rightarrow z_1 & \rightarrow dis & \rightarrow es & \rightarrow y_3 & \rightarrow e \\
\overline{d} & \overline{z_1} & \overline{dis} & \overline{es} & \overline{y_3} & \overline{e}
\end{array}
$$

$|\!\!\longrightarrow\!\longrightarrow$

$$
\begin{array}{cccccc}
e & \rightarrow z_2 & \rightarrow x_3 & \rightarrow f & \rightarrow y_4 & \rightarrow fis \\
\overline{e} & \overline{z_2} & \overline{x_3} & \overline{f} & \overline{y_4} & \overline{fis}
\end{array}
$$

$s_3 \;|\!\!\longrightarrow\!\longrightarrow\!\!\!|\!\!\longrightarrow\!$

$$
\begin{array}{cccccc}
fis & \rightarrow ges & \rightarrow x_4 & \rightarrow g & \rightarrow y_5 & \rightarrow gis \\
\overline{fis} & \overline{ges} & \overline{x_4} & \overline{g} & \overline{y_5} & \overline{gis}
\end{array}
$$

$\longrightarrow\!\!\!|\!\!\longrightarrow$

$$
\begin{array}{cccccc}
gis & \rightarrow as & \rightarrow x_5 & \rightarrow a & \rightarrow y_6 & \rightarrow ais \\
\overline{gis} & \overline{as} & \overline{x_5} & \overline{a} & \overline{y_6} & \overline{e}
\end{array}
$$

$\longrightarrow\!\!\!|\!\!\longrightarrow$

$$
\begin{array}{cccccc}
ais & \rightarrow b & \rightarrow x_6 & \rightarrow h & \rightarrow w_1 & \rightarrow w_2 \\
\overline{ais} & \overline{b} & \overline{x_6} & \overline{h} & \overline{w_1} & \overline{w_2}
\end{array}
$$

$\longrightarrow\!\!\!|\!\!\longrightarrow$

$$
\begin{array}{ccccc}
w_2 & \rightarrow c^1 ; 31 & \rightarrow y_1^1 & \rightarrow cis^1 & \rightarrow des^1 & \rightarrow d^1 \\
\overline{w_2} & \overline{c^1} ; 7 & \overline{y_1^1} & \overline{cis^1} & \overline{des^1} & \overline{d^1}
\end{array}
$$

$\longrightarrow$

That is, the relationship between the two parameters $\delta/2\pi$ and $\xi/2\pi$ is given by the logarithmic law

$$\xi/2\pi = \log_2(1 + \delta/2\pi), \tag{9.5}$$

which represents the inverse function to the exponential law, Eq. (9.2), see also refs. [3], [5]. The two functions, Eqs. (9.2) and (9.5), being

Table 9.1. Interval Vectors, Cents, and Interval Factors for the 31-Tone Three-Dimensional Musical System

$l((n,m,r),(0,0,0)) = 1{,}200\,n + 701.955\,m + 884.358\,708\,r$

$S_1^{-1} = (3,0,-4) = 62.565\,224,$
$S_1^2 S_2 = (-5,1,6) = 8.107\,248,$
$p^{-1} = (1,-3,1) = -21.506\,288$

$S_2 = (1,1,-2) = 133.237\,584$
$S_1^{-1}p^{-1} = (4,-3,-3) = 41.058\,876$

$w_2^{-1} = S_1 p = (-4,3,3) = -41.058\,876$; $S_1 p$; S_1^{-1}

c	$= (0,0,0)$	$=$	0.0	; 1	; S_1^{-1}	
y_1	$= (3,0,-4)$	$=$	$62.565\,224$	; S_1^{-1}	; $S_1^2 S_2$	
cis	$= (-2,1,2)$	$=$	$70.672\,416$	; $S_1 S_2$	; S_1^{-1}	
des	$= (1,1,-2)$	$=$	$133.237\,584$	; S_2	; $S_1^2 S_2$	
y_2	$= (-4,2,4)$	$=$	$141.344\,832$	; $S_1^2 S_2^2$	; S_1^{-1}	
d	$= (-1,2,0)$	$=$	$203.910\,000$	; $S_1 S_2^2$	; S_1^{-1}	; $S_1 S_2^2$
z_1	$= (2,2,-4)$	$=$	$266.475\,168$	; S_2^2	; $S_1^2 S_2$	
dis	$= (-3,3,2)$	$=$	$274.582\,416$	; $S_1^2 S_2^3$	; $S_1^{-1}p^{-1}$	
es	$= (1,0,-1)$	$=$	$315.641\,292$	; $S_1 S_2^3 p^{-1}$	; $S_1^2 S_2$	
y_3	$= (-4,1,5)$	$=$	$323.748\,540$	; $S_1^3 S_2^4 p^{-1}$	; S_1^{-1}	
e	$= (-1,1,1)$	$=$	$386.313\,708$	; $S_1^2 S_2^4 p^{-1}$	; $S_1^{-1}p^{-1}$	; $S_1 S_2^2 p^{-1}$
z_2	$= (3,-2,-2)$	$=$	$427.372\,584$	; $S_1 S_2^4 p^{-2}$	; $S_1^2 S_2$	
x_3	$= (-2,-1,4)$	$=$	$435.479\,832$	; $S_1^3 S_2^5 p^{-2}$	; S_1^{-1}	
f	$= (1,-1,0)$	$=$	$498.045\,000$	; $S_1^2 S_2^5 p^{-2}$	; $S_1^2 S_2$	
y_4	$= (-4,0,6)$	$=$	$506.152\,248$	; $S_1^4 S_2^6 p^{-2}$	; S_1^{-1}	
fis	$= (-1,0,2)$	$=$	$568.717\,416$	; $S_1^3 S_2^6 p^{-2}$	; S_1^{-1}	; $S_1 S_2^2 p^{-1}$
ges	$= (2,0,-2)$	$=$	$631.282\,584$	; $S_1^2 S_2^6 p^{-2}$	; $S_1^2 S_2$	
x_4	$= (-3,1,4)$	$=$	$639.389\,832$	; $S_1^4 S_2^7 p^{-2}$	; S_1^{-1}	
g	$= (0,1,0)$	$=$	$701.955\,000$	; $S_1^3 S_2^7 p^{-2}$	; $S_1^2 S_2$	
y_5	$= (-5,2,6)$	$=$	$710.062\,218$	; $S_1^5 S_2^8 p^{-2}$	; S_1^{-1}	
gis	$= (-2,2,2)$	$=$	$787.796\,240$	; $S_1^4 S_2^8 p^{-2}$	; $S_1^{-1}p^{-1}$	; $S_1 S_2^2$
as	$= (2,-1,-1)$	$=$	$813.686\,297$	; $S_1^3 S_2^8 p^{-3}$	; $S_1^2 S_2$	
x_5	$= (-3,0,5)$	$=$	$821.793\,540$	; $S_1^5 S_2^9 p^{-3}$	; S_1^{-1}	
a	$= (0,0,1)$	$=$	$884.358\,708$	; $S_1^4 S_2^9 p^{-3}$	; $S_1^2 S_2$	
y_6	$= (-5,1,7)$	$=$	$892.465\,956$	; $S_1^6 S_2^{10} p^{-3}$	; S_1^{-1}	
ais	$= (-2,1,3)$	$=$	$955.031\,124$	; $S_1^5 S_2^{10} p^{-3}$	; S_1^{-1}	; $S_1 S_2^2 p^{-1}$
b	$= (1,1,-1)$	$= 1{,}017.596\,29$		; $S_1^4 S_2^{10} p^{-3}$	; $S_1^2 S_2$	
x_6	$= (-4,2,5)$	$= 1{,}025.703\,54$		; $S_1^6 S_2^{11} p^{-3}$	; S_1^{-1}	

Table 9.1 (*continued*)

h	$= (-1, 2, 1)$	$= 1{,}088.268\ 71$	$; S_1^5 S_2^{11} p^{-3}$	$; S_1^2 S_2$	
w_1	$= (-6, 3, 7)$	$= 1{,}096.375\ 96$	$; S_1^7 S_2^{12} p^{-3}$	$; S_1^{-1}$	
w_2	$= (-3, 3, 3)$	$= 1{,}158.941\ 12$	$; S_1^6 S_2^{12} p^{-3}$	$; S_1^{-1} p^{-1}$	$; S_1 S_2^2$
c^1	$= (1, 0, 0)$	$= 1{,}200.000\ 00$	$; S_1^5 S_2^{12} p^{-4} = 2$	$; S_1^{-1} p^{-1}$	

$|(S_1 S_2^2)|(S_1 S_2^2\ p^{-1})|(S_2 p^{-1})|(S_1 S_2^2)|(S_1 S_2^2 p^{-1})|(S_1 S_2^2)|(S_2 p^{-1})|$;
$(S_1 S_2^2)^3 (S_1 S_2^2 p^{-1})^2 (S_2 p^{-1})^2 = 2$, the 7-tone natural diatonic scale

Table 9.2. Interval Vectors, Cents, and Interval Factors for the 29-Tone Two-Dimensional Musical System

$I((n, m, 0), (0, 0, 0)) = 1{,}200\ n + 701.955\ m$

$s_1 = (-7, 12, 0)\ = 23.46$ cent
$s_2 = (10, -17, 0) = 66.765$ cent
$p\ = (-1, 3, -1)\ = 21.506\ 288$ cent

$\bar{y}_1$	$= (7, -12, 0)$	$= y_1 - 4p$	$= -23.46$	$; s_1^{-1}$
$\bar{c}$	$= (0, 0, 0)$	$= c$	$= 0$	$; 1$
$\bar{w}_2^{-1}$	$= (-7, 12, 0)$	$= w_2^{-1} + 4p$	$= 23.46$	$; s_1$
des	$= (3, -5, 0)$	$= des - 2p$	$= 90.225$	$; s_1 s_2$
$\bar{c}is$	$= (-4, 7, 0)$	$= cis + 2p$	$= 113.685$	$; s_1^2 s_2$
$\bar{z}_1$	$= (6, -10, 0)$	$= z_1 - 4p$	$= 180.45$	$; s_1^2 s_2^2$
$\bar{d}$	$= (-1, 2, 0)$	$= d$	$= 203.91$	$; s_1^3 s_2^2 = t$
$\bar{y}_2$	$= (-8, 14, 0)$	$= y_2 + 4p$	$= 227.37$	$; s_1^4 s_2^2$
$\bar{e}s$	$= (2, -3, 0)$	$= es - p$	$= 294.135$	$; s_1^4 s_2^3$
dis	$= (-5, 9, 0)$	$= dis + 2p$	$= 317.595$	$; s_1^5 s_2^3$
$\bar{z}_2$	$= (5, -8, 0)$	$= z_2 - 2p$	$= 384.36$	$; s_1^5 s_2^4$
$\bar{e}$	$= (-2, 4, 0)$	$= e + p$	$= 407.82$	$; s_1^6 s_2^4 = t^2$
$\bar{y}_3$	$= (-9, 16, 0)$	$= y_3 + 5p$	$= 431.28$	$; s_1^7 s_2^4$
$\bar{f}$	$= (1, -1, 0)$	$= f$	$= 498.054$	$; s_1^7 s_2^5$
$\bar{x}_3$	$= (-6, 11, 0)$	$= x_3 + 4p$	$= 521.505$	$; s_1^8 s_2^5$
$\bar{g}es$	$= (4, -6, 0)$	$= ges - 2p$	$= 588.27$	$; s_1^8 s_2^6$
$\bar{f}is$	$= (-3, 6, 0)$	$= fis + 2p$	$= 611.73$	$; s_1^9 s_2^6 = t^3$
$\bar{y}_4$	$= (-10, 18, 0)$	$= y_4 + 6p$	$= 635.19$	$; s_1^{10} s_2^6$
$\bar{g}$	$= (0, 1, 0)$	$= g$	$= 701.955$	$; s_1^{10} s_2^7$
$\bar{x}_4$	$= (-7, 13, 0)$	$= x_4 + 4p$	$= 725.415$	$; s_1^{11} s_2^7$
$\bar{a}s$	$= (3, -4, 0)$	$= as - p$	$= 792.18$	$; s_1^{11} s_2^8$

(*continued*)

Table 9.2 (*continued*)

$\bar{g}is$	$=(-4,8,0)$	$=as+2p$	$=815.64$	$;s_1^{12}s_2^{8}=t^4$
$\bar{y}_5$	$=(-11,20,0)$	$=y_5+6p$	$=839.1$	$;s_1^{13}s_2^{8}$
$\bar{a}$	$=(-1,3,0)$	$=a+p$	$=905.865$	$;s_1^{13}s_2^{9}$
$\bar{x}_5$	$=(-8,15,0)$	$=x_5+5p$	$=929.325$	$;s_1^{14}s_2^{9}$
$\bar{b}$	$=(2,-2,0)$	$=b-p$	$=996.09$	$;s_1^{14}s_2^{10}$
$\bar{a}is$	$=(-5,10,0)$	$=ais+3p$	$=1{,}019.55$	$;s_1^{15}s_2^{10}=t^5$
$\bar{y}_6$	$=(-12,22,0)$	$=y_6+7p$	$=1{,}043.01$	$;s_1^{16}s_2^{10}$
$\bar{h}$	$=(-2,5,0)$	$=h+p$	$=1{,}109.775$	$;s_1^{16}s_2^{11}$
$\bar{x}_6$	$=(-9,17,0)$	$=x_6+5p$	$=1{,}133.235$	$;s_1^{17}s_2^{11}$
$\bar{c}^1$	$=(1,0,0)$	$=c^1$	$=1{,}200$	$;s_1^{17}s_2^{12}=2$
$\bar{w}_2$	$=(-6,12,0)$	$=w_2+3p$	$=1{,}223.46$	$;s_1^{18}s_2^{12}$
$\bar{w}_1=\bar{d}es^1$	$=(-13,24,0)$	$=w_1+7p$	$=1{,}246.92$	$;s_1^{19}s_2^{12}$
	$t=(s_1^3s_2^2),$	$/t/t/t/t/t/ts_1^{-1}$		

The sequence of intervals given below represents a 22/23-tone musical system given in 2-dimensional lattice space:

$$|(s_1s_2)s_1s_2s_1|(s_1s_2)s_1s_2s_1|(s_1s_2)s_1s_2s_1|(s_1s_2)s_1s_2s_1|s_2s_1(s_1s_2)s_1|s_2s_1(s_1s_2)$$

inverse to each other, carry the same information. A physical law, represented by a mathematical function, needs to be unique. This requirement of uniqueness of a functional relationship led to the postulate that "musical lattice tones" are defined for the parameter values $\delta/2\pi=\xi/2\pi=0$ and 1 only, and that this requirement holds for each of the three octave systems based upon the frequency ratios $(2/1)$, $(3/2)$ and $(5/3)$. This then led to the introduction of the 3-dimensional scaled lattice space for musical tones, see also ref. [3].

The ratio of two distinct frequency ratios ν_1/ν_0 and ν_2/ν_0 (i.e. two distinct musical tones) is given by

$$(\nu_2/\nu_0)/(\nu_1/\nu_0)=\nu_2/\nu_1=(1+\delta_2/2\pi)/(1+\delta_1/2\pi)$$
$$=2^{\xi_2/2\pi}/2^{\xi_1/2\pi}=2^{(\xi_2/2\pi)-(\xi_1/2\pi)}. \tag{9.6}$$

Taking the logarithm with base 2 one obtains

$$\log_2(\nu_2/\nu_1)=\log_2(1+\delta_2/2\pi)-\log_2(1+\delta_1/2\pi)$$
$$=(\xi_2/2\pi)-(\xi_1/2\pi)=l. \tag{9.7}$$

That is, the ratio of the two frequencies ν_2/ν_1 (the ratio of the two musical tones (ν_1/ν_0) and (ν_2/ν_0)) is mapped upon the distance l,

$$0\le l=(\xi_2/2\pi)-(\xi_1/2\pi)\le 1.$$

If $(\xi_2/2\pi) - (\xi_1/2\pi) = 0$, the distance is zero and the two parameter values obviously represent the same tone. If $(\xi_2/2\pi) - (\xi_1/2\pi) = 1$, then the two tones differ by an octave.

The distances obtained in this manner are unhandy small numbers. Thus, by convention, the distance formula is renormalized by the factor 1,200 and is given the name cent (out of 1,200 cent),

$$l(\text{cent}) = 1{,}200\log_2(\nu_2/\nu_1) = 1{,}200(1/\log_{10}2)\log_{10}(\nu_2/\nu_1)$$
$$= 3{,}986.313\,71\log_{10}(\nu_2/\nu_1). \tag{9.8}$$

This is the standard formula for the distance (interval) between two musical tones in terms of cent.

Given two musical tones in the form of lattice points,

$$(\nu_2/\nu_0) = (n, m, r) \quad \text{and} \quad (\nu_1/\nu_0) = (k, l, u), \qquad n, m, r, k, l, u \text{ integers,}$$

the frequency ratio of the two tones is obtained as

$$\nu_2/\nu_1 = (2/1)^n(3/2)^m(5/3)^r/(2/1)^k(3/2)^l(5/3)^u$$
$$= (n - k, m - l, r - u).$$

Thus it follows

$$\log_2(\nu_2/\nu_1) = \log_2(2/1)^{n-k} + \log_2(3/2)^{m-l} + \log_2(5/3)^{r-u}$$
$$= n - k + (m - l)1.584\,962\,50 + (r - u)2.321\,928\,09.$$

The cent are then given by the formula

$$l(\text{cent}) = 1{,}200(n - k) + 701.9555\,00(m - l) + 884.358\,71(r - u). \tag{9.9}$$

It is thus seen that the cents for the distance between two musical tones are given by the sum of the cents along the (2/1)-based octave, the (3/2)-based octave and the (5/3)-based octave, respectively. The distance (in cent) between two musical tones is thus expressed as a *linear equation in terms of three discrete parameters*.

10. Determination of Tones – the Three-Dimensional 116-Tone System

In this section the Table of Tones, as given in [2], pp. 796–801, is discussed. It will be shown that, by adding a few tones to this list, a structured lattice tone system of 116 tones is obtained which is based upon three intervals (vectors – lattice points) λ, μ, ρ only.

Table 10.1 Inventory of Tones: List of Tones and Tonal System Structure for the Interval $c - d$

Ordered interval sequences for (1): inventory of 116 tones in terms of $\{\lambda, \mu, \rho\}$, (2): 31-tone system in terms of $\{S_1^{-1}, S_1^2 S_2, p\}$, (3): Carnatic 23-tone system in terms of $\{\alpha, p, \gamma\}$ and subsystems, for the interval $d/c = T_1 = \lambda^{11}\mu^7\rho^2$, $(c = 1)$.

Frequency ratio ν_1/ν_2 (numerical value associated with the lattice vector (n, m, r): $\nu_1/\nu_2 = (2/1)^n (3/2)^m (5/3)^r = (n, m, r)$; i.e. multiplication of frequency ratios is equivalent to addition of their vectors. $(1, 0, 0) = O$ (Octave) $= c^1/c = (2/1)$. $(-1, 1, 1) = T$ (Terz) $= e/c = (5/4)$, $(0, 1, 0) = Q$ (Quint) $= g/c = (3/2)$, $(0, 0, 1) = S$ (Sixth) $= a/c = (5/3)$, with $S = OT/Q$, for the numerical values, or equivalently $(0, 0, 1) = ((1, 0, 0) + (-1, 1, 1)) - (0, 1, 0)$, for the vector components $(n, m, r) = (2/1)^n (3/2)^m (5/3)^r$.

	#1	#2	#3	#4	#5	#6	#7	#8
#1	c	$(0, 0, 0)$	1	λ	1	S_1^{-1}	1	α
#2	λ	$(-6, 9, 1)$	λ	μ				
#3	$S_1^2 S_2$	$(-5, 1, 6)$						
#4	μ	$(11, -15, -3)$						
#5	$\lambda\mu$	$(5, -6, -2)$	$\lambda\mu$	λ				
#6	p	$(-1, 3, -1)$	$\lambda^2\mu$	λ				
#7	s_1	$(-7, 12, 0)$	$\lambda^3\mu$	μ				
#8	$\rho = S_1^2 S_2 p$	$(-6, 4, 5)$						
#9	$S_1^{-1} p^{-1}$	$(4, -3, -3)$	$\lambda^3\mu^2$	ρ				
#10	p^2	$(-2, 6, -2)$						
#11	$S_1 S_2 p^{-1} = \lambda\mu\rho$	$(-1, -2, 3)$						
#12	$y_1 = S_1^{-1}$	$(3, 0, -4)$			S_1^{-1}	$S_1^2 S_2$		
#13	s_2	$(10, -17, 0)$						
#14	$cis = \gamma = S_1 S_2 = $ ♯	$(-2, 1, 2)$	$\lambda^3\mu^2\rho$	λ	$S_1 S_2$	S_1^{-1}		
#15		$(-8, 10, 3)$	$\lambda^4\mu^2\rho$	μ				
#16	$\alpha = s_1 s_2 = Sp^{-1} = s_3$	$(3, -5, 0)$	$\lambda^4\mu^3\rho$	λ			α	p

#17	$S_1 S_2 p$	$(-3, 4, 1)$	$\lambda^5 \mu^3 \rho$	λ				
#18		$(-9, 13, 2)$	$\lambda^6 \mu^3 \rho$	μ				γ
#19	$S = S_2 p^{-1} = \alpha p = b_2$	$(2, -2, -1)$	$\lambda^6 \mu^4 \rho$	λ			αp	
#20	$p\gamma p = s_1^2 s_2$	$(-4, 7, 0)$	$\lambda^7 \mu^4 \rho$	λ				
#21		$(-10, 16, 1)$	$\lambda^8 \mu^4 \rho$	μ				
#22		$(7, -8, -3)$						
#23	$des = S_2 = b_1$	$(1, 1, -2)$	$\lambda^8 \mu^5 \rho$	ρ	S_2	$S_1^2 S_2$		
#24	$y_2 = S_1^2 S_2^2$	$(-4, 2, 4)$			$S_1^2 S_2^2$	S_1^{-1}		
#25	$T_2 p^{-1}$	$(1, -4, 2)$						
#26	λT_1	$(-5, 5, 3)$	$\lambda^8 \mu^5 \rho^2$	μ				
#27	ρT_1	$(6, -10, 0)$	$\lambda^8 \mu^6 \rho^2$	λ				
#28	$T_2 = T_1 p^{-1} = \alpha p \gamma$	$(0, -1, 1)$	$\lambda^9 \mu^6 \rho^2$	λ			$\alpha p \gamma$	p
#29	λT_2	$(-6, 8, 2)$	$\lambda^{10} \mu^6 \rho^2$	μ				
#30	ρT_2	$(5, -7, -1)$	$\lambda^{10} \mu^7 \rho^2$	λ				
#31	$d = T_1 = S_1 S_2^2$	$(-1, 2, 0)$	$\lambda^{11} \mu^7 \rho^2$	λ	$S_1 S_2^2$	S_1^{-1}	$\alpha p \gamma p$	α

(continued)

Table 10.1 (*continued*)

Column #1: algebraic designation of tone; column #2: lattice point (vector) corresponding to tone; column #3: 116-tone sequence: build up of the tones λ, μ, ρ; column #4: the intervals between successive tones of the sequence; column #5: 31-tone sequence, build up of the tones S_1^{-1}, $S_1^2 S_2$, p; column #6: intervals between two successive tones of the sequence; columns #7 and #8: 23-tone Carnatic system based upon the tones α, p, γ; column #9: notation for the tones in the O, T, Q-basis, and in the O, Q, S-basis used in this article; column #10: list of names for the intervals/tones, ref. [8]; column #11: frequency ratio of tone; ν/c; column #12: interval I of tones ν/c in cents, $I((n,m,r).\,(0,0,0)) = 1{,}200\,n + 701.955\,m + 884.358\,708\,r$

	#9	#10	#11	#12
#1	prime	Unison	1.0	0.0
#2	$T(8Q)/(5O) = (9Q)S/(6O)$	Schisma	1.001 129 15	1.953 708
#3	$(6T\,O)/(5Q) = Q(6S)/(5O)$	Kleisma	1.004 693 93	8.107 248
#4	$(8O)/(3T)(12Q) = (11O)/(15Q)(3S)$	Diaschisma-schisma	1.010 217 34	17.598 876
#5	$(3O)/(2T)(4Q) = (5O)/(6Q)(2S)$	Diaschisma	1.011 358 02	19.552 584
#6	$(4Q)/T(2O) = (3Q)/OS$	Comma syntonum Pythagorean vector p	1.0125	21.506 292
#7	$(12Q)/(7O)$	Comma of Pythagoras	1.013 643 26	23.46
#8	$(5T)/OQ = (4Q)(5S)/(6O)$	Small diesis	1.017 252 6	29.613 54
#9	$O/(3T) = (4O)/(3Q)(3S)$	Diesis minor	1.024	41.058 876
#10	$(8Q)/(2T)(4O) = (6Q)/(2O)(2S)$	p^2	1.025 156 25	43.012 584
#11	$(3T)(2O)/(5Q) = (3S)/O(2Q)$	Maximal diesis	1.028 806 58	49.166 124
#12	$(4Q)/O(4T) = (3O)/(4S)$	Major diesis	1.0368	62.565 168
#13	$(10O)/(17Q)$	Pyth. double dimin. 3rd	1.039 318 25	66.765
#14	$(2T)/Q = Q(2S)/(2O)$	Minor chroma	1.041 666 67	70.672 416
#15	$(3T)(7Q)/(5O) = (10Q)(3S)/(8O)$		1.042 842 86	72.626 124
#16	$(3O)/(5Q)$	Limma	1.053 497 94	90.225

#	Formula	Description	Value	Cents
#17	$T(3Q)/(2O) = (4Q)S/(3O)$	Major chroma	1.054 687 5	92.178 708
#18	$(11Q)(2T)/(7O)(13Q)(2S)/(9O)$		1.056	94.132 416
#19	$O/TQ = (2O)/(2Q)S$	Leading tone step, semitone	1.066 666 67	111.731 292
#20	$(7Q)/(4O)$	Apotome	1.067 871 09	113.685
#21	$T(15Q)/(9O) = (16Q)S/(10O)$		1.069	115.638 708
#22	$(4O)/(5Q)(3T) = (7O)/(8Q)(3S)$		1.078 781 89	131.283 876
#23	$(3Q)/(2T)O = OQ/(2S)$	Large limma	1.08	133.237 58
#24	$(4T)/(2Q) = (2Q)(4S)/(4O)$	BP great semitone	1.085 069	141.344 83
#25	$(2T)(3O)/(6Q) = O(2S)/(4Q)$	Grave whole tone	1.097 393 69	160.897 416
#26	$(3T)/(2Q)(2O) = (5Q)(3S)/(5O)$	Double augm. prime	1.098 632 81	162.851 124
#27	$(6O)/(10Q)$	$(\text{Limma})^2$	1.109 857 91	180.45
#28	$TO(2Q) = S/Q$	Minor whole tone	1.111 111 11	182.403 708
#29	$(2T)(6Q)/(4O) = (8Q)(2S)/(6O)$		1.112 365 72	184.357 416
#30	$(4O)/T(6Q) = (5O)/(7Q)S$		1.123 731 14	201.956 292
#31	$(2Q)/O$	Major whole tone	1.125	203.91

The system of tones, as presented in ref. [2], uses the (musically motivated) basis

$$
\begin{aligned}
(1,0,0) \quad &= O \text{ (Octave)} \quad = 2/1 \quad = c^1/c, \\
(-1,1,1) \quad &= T \text{ (Terz)} \quad = 5/4 \quad = e/c, \\
(0,1,0) \quad &= Q \text{ (Quint)} \quad = 3/2 \quad = g/c,
\end{aligned} \tag{10.1}
$$

while the (mathematically motivated) basis used in this article is

$$
\begin{aligned}
(1,0,0) \quad &= O \text{ (Octave)} \quad = 2/1 \quad = c^1/c, \\
(0,1,0) \quad &= Q \text{ (Quint)} \quad = 3/2 \quad = g/c, \\
(0,0,1) \quad &= S \text{ (Sixth)} \quad = 5/3 \quad = a/c.
\end{aligned} \tag{10.2}
$$

The two bases are mathematically equivalent,

$$
S = OT/Q. \tag{10.3}
$$

Introducing the intervals (tones)

$$
\begin{aligned}
\lambda \quad &= (-6,9,1) \quad &= T(8Q)/(5O), \\
\mu \quad &= (11,-15,-3) \quad &= (8O)/(3T)(12Q), \\
\rho \quad &= (-6,4,5) \quad &= (5T)/OQ,
\end{aligned} \tag{10.4}
$$

as basis intervals, the closure condition

$$
k_1 \lambda + k_2 \mu + k_3 \rho = (1,0,0) \tag{10.5}
$$

is satisfied for

$$
k_1 = 63, \qquad k_2 = 41, \qquad k_3 = 12, \qquad N = 116. \tag{10.6}
$$

Thus a total of 116 tones is obtained.

For purposes of illustration the list of tones for the interval $d/c - c/c$ is given in detail in Table 10.1. For the interval $c - d$ holds

$$
k_1 \lambda + k_2 \mu + k_3 \rho = (-1,2,0), \quad k_1 = 11, \quad k_2 = 7, \quad k_3 = 2, \quad N = 20 \tag{10.7}
$$

and thus the tone interval $c - d$ contains 20 tones. The 30 tones contained in Table 10.1 are then obtained by adding to the list the tones $S_1^2 S_2 = (-5,1,6)$, $\mu = (11,-15,-3)$, $\rho = (-6,4,5)$, $p^2 = (-2,6,-2)$, $(-8,15,-1)$, $S_1^{-1} = (3,0,-4)$, $s_2 = (10,-17,0)$, $(-9,13,2)$, $(7,-8,-3)$, $y_2 = (-4,2,-4)$. Most of these added tones are base intervals for other musical tone systems.

The 20-tone tonal *system* for the interval $c - d$, that is, for the interval T_1, is then given, in sequential order, by

$$
\begin{aligned}
T_1 &= /\lambda\mu\lambda\lambda\mu\rho\lambda\mu\lambda\lambda\mu\lambda\lambda\mu\rho\mu\lambda\lambda\mu\lambda/ \\
&= S_3 + S - p.
\end{aligned} \tag{10.8}
$$

Similarly, for the interval T_2, the tone interval $d - e$, a 17-tone sequence is obtained as

$$T_2 = /\lambda\mu\lambda\lambda\mu\rho\lambda\mu\lambda\lambda\mu\lambda\lambda\mu\rho\mu\lambda/$$
$$= S_3 + S \qquad (10.9)$$

and for the interval S, the tone interval $e - f$, an 11-tone sequence is obtained as

$$S = /\lambda\mu\lambda\lambda\mu\rho\mu\lambda\lambda\mu\lambda/. \qquad (10.10)$$

This then yields a 116-tonal system for the octave by means of the (ordered) sequence

$$/T_1/T_2/S/T_1/T_2/T_1/S/ \qquad (10.11)$$

having the octave property

$$\lambda^{63}\mu^{41}\rho^{12} = 2, \qquad N = 116. \qquad (10.12)$$

It needs to be noted that Eqs. (10.8)–(10.10) represent only one particular choice for the sequence of the intervals. This choice for the sequence of intervals has been motivated by the aim to include in a regular manner, as many as possible, of the tones of the List of Tones of [2], and keeping the introduction of new tones to a minimum.

The 116-tone system contains, as sub-systems (Table 10.1),

(a) the tones of the 31-tone system,
(b) the standard chromatic major and minor scales,
(c) the natural diatonic scale,
(d) the tone scale of the ancient Greek Lyre,
(e) the tones of the hypothetical 22/23-tone Carnatic scale,
(f) and the subsystems/scales of all these systems,
(g) the 2-dimensional tonal lattices in the Pythagorean plane by means of projection along the Pythagorean vector into the Pythagorean plane.

Column #1 of Table 10.1 lists the mathematical-musical designation of the tones. Column #2 lists the tone by its lattice point coordinates (n, m, r). Column #3 indicates the cumulative build up of the tones based upon the basis λ, μ, ρ. The λ, μ, ρ may be looked upon as frequency ratios or as vectors (for example, the entry $\lambda^3\mu^2$, can be read as $3\lambda + 2\mu = 3(-6, 9, 1) + 2(11, -15, -3) = (-18, 27, 3) + (22, -30, -6) = (4, -3, -3))$. Column #4 shows the interval factor/vector between two neighboring tones. Columns #5 and #6 refer to the 31-tone system based upon the base elements $S_1^{-1} = (3, 0, -4)$, $S_1^2 S_2 = (-5, 1, 6)$ and $p = (-1, 3, -1)$. Column #5

Table 10.2. Relationships Between Theoretical (3-Dimensional) Carnatic System, Chromatic Scale and 12-Tone (2-Dimensional) Pythagorean Scale

Columns #1 to #4: The theoretical 3-dimensional Carnatic tone system.
Column #1: Frequency ratios of the theoretical 3-dimensional Carnatic tone system.
Column #2: The lattice vectors corresponding to the frequency ratios given in #1.
Column #3: The interval-vectors (adding #3 to #2 results in the next tone).
Column #4: Decimal value of the tone.
Columns #5 and #7: The chromatic tone related to the Carnatic tone by the vector p.
Column #6: The chromatic tone #6 projected into the Pythagorean plane.
Columns #7 to #9: The 12-tone 2-dimensional Pythagorean scale.
Column #10: The interval vectors for the 7-tone diatonic Pythagorean scale, corresponding to the set of 7 Pythagorean diatonic tones of #7.

	#1	#2	#3	#4	#5	#6
#1	1/1	(0,0,0)	$\alpha = (3,-5,0) = s = s_1 s_2 = S_2 p^{-2}$	1.0	(0,0,0)	c
#2	256/243	(3,−5,0)	$p = (-1,3,-1)$	1.053 498	$(1,1,-2) - 2p = des - 2p$	des
#3	16/15	(2,−2,−1)	$\gamma = (-2,1,2) = S_1 S_2 = s_1 s_2 p^{-2}$	1.066 667	$(1,1,-2) - p = des - p$	des
#4	10/9	(0,−1,1)	p	1.111 111	$(-1,2,0) - p = d - p$	d
#5	9/8	(−1,2,0)	$\alpha = Sp^{-1}$	1.125	$(-1,2,0) = d$	d
#6	32/27	(2,−3,0)	p	1.185 185	$(1,0,-1) - p = es - p$	es
#7	6/5	(1,0,−1)	γ	1.2	$(1,0,-1) = es$	es
#8	5/4	(−1,1,1)	p	1.25	$(-1,1,1)$	e
#9	81/64	(−2,4,0)	$\alpha = Sp^{-1}$	1.265 625	$(-1,1,1) + p = e + p$	e
#10	4/3	(1,−1,0)	p	1.333 333	$(1,-1,0) = f$	f
#11	27/20	(0,2,−1)	α to (3,−3,−1)	1.35	$(1,-1,0) + p = f + p$	f
#12	25/18	(−1,0,2)	α	1.388 889	$(-1,0,2) = fis$	fis
#13	45/32	(−2,3,1)	p	1.406 25	$(-1,0,2) + p = fis + p$	fis

#14	64/45	$(3, -3, -1)$	p	1.422 222	$(2, 0, -2) - p = ges - p$	ges
#15	36/25	$(2, 0, -2)$	$(\gamma p^{-1}) = (-1, -2, 3)$	1.44	$(2, 0, -2) = ges$	ges
#16	40/27	$(1, -2, 1)$	p	1.481 481	$(0, 1, 0) = g - p$	g
#17	3/2	$(0, 1, 0)$	α	1.5	$(0, 1, 0) = g$	g
#18	128/81	$(3, -4, 0)$	p	1.580 247	$(2, -1, -1) - p = as - p$	as
#19	8/5	$(2, -1, -1)$	γ	1.6	$(2, -1, -1) = as$	as
#20	5/3	$(0, 0, 1)$	p	1.666 667	$(0, 0, 1) = a$	a
#21	27/16	$(1, -3, 0)$	α	1.687 5	$(0, 0, 1) + p = a + p$	a
#22	16/9	$(2, -2, 0)$	p	1.777 778	$(1, 1, -1) - p = b - p$	b
#23	9/5	$(1, 1, -1)$	γ	1.8	$(1, 1, -1) = b$	b
#24	15/8	$(-1, 2, 1)$	p	1.875	$(-1, 2, 1) = h$	h
#25	243/128	$(-2, 5, 0)$	α	1.898 437	$(-1, 2, 1) + p = h + p$	h
#26	2/1	$(1, 0, 0)$		2.0	$(1, 0, 0) = c^1$	c^1

(continued)

Table 10.2 (continued)

Theoretical 3-dimensional Carnatic musical system: Interval vectors $\{\alpha = (3, -5, 0),\ p = (-1, 3, -1),\ \gamma = (-2, 1, 2),\ (\gamma p^{-1}) = (-1, -2, 3)\},\ \alpha^7 p^{11} \gamma^4 (\gamma p^{-1}) = 2$, 12-tone Pythagorean scale: $\{\alpha = (3, -5, 0),\ (p\gamma p) = (-4, 7, 0)\},\ \alpha^7 (p\gamma p)^5 = 2$, 7-tone diatonic Pythagorean scale: $\{(\alpha p\gamma p) = (-1, 2, 0),\ \alpha = (3, -5, 0)\},\ (\alpha p\gamma p)^5 \alpha^2 = 2$.

Relationships to other tonal systems: $\alpha p\gamma p = T_1 = S_1 S_2^2 = s_1^3 s_2^2,\ \alpha = Sp^{-1} = S_2 p^{-2} = s = s_1 s_2,\ \gamma = S_1 S_2 = s_1^2 s_2 p^{-2},\ \gamma p^{-1} = S_1 S_2^2 p^{-1},\ s_1 s_2 = s_3,\ S_1 S_2 = s,\ S_1 S_2 = S_3,\ p\gamma p = \beta = s_1^2 s_2 = (-4, 7, 0)$, Pythagorean comma: $\beta\alpha^{-1} = (-7, 12, 0)$.

#6	#7	#8	#9	#10	#11
c	$\bar{c} = c = (0,0,0)$	$\alpha = (3, -5, 0) = s = s_1 s_2$	1/1	$T_1 = (\alpha p\gamma p)$	1.0
des					
des	$\bar{d}es = (3, -5, 0)$	$(p\gamma p) = (-4, 7, 0) = s_1^2 s_2$	256/143		1.053 498
d					
d	$\bar{d} = d = (-1, 2, 0)$	α	9/8	$T_1 = (\alpha p\gamma p)$	1.125
es					
es	$\bar{e}s = (2, -3, 0)$	$(p\gamma p)$	32/27		1.185 185
e					
e	$\bar{e} = (-2, 4, 0)$	α	81/64	$\alpha = (Sp^{-1})$	1.265 625
f	$\bar{f} = f = (1, -1, 0)$	α to $\bar{g}es$; $(p\gamma p)$ to $\bar{f}is$	4/3	$T_1 = (\alpha p\gamma p)$ to $\bar{g}es$	1.333 333
f					
fis					
fis	$\bar{f}is = (-3, 6, 0)$	$(7, -12, 0) = (\beta^{-1}\alpha)$ to $\bar{g}es$; α to g	729/512		1.423 828

ges					
ges	$\bar{g}es = (4,-6,0)$	$(p\gamma p) = (-4,7,0) = s_1^2\,s_2$	1,024/729		1.404 664
g					
g	$\bar{g} = g = (0,1,0)$	α	3/2	$T_1 = (\alpha p\gamma p)$	1.5
as					
as	$\bar{a}s = (3,-4,0)$	$(p\gamma p)$	128/81		1.580 247
a					
a	$\bar{a} = (-1,3,0)$	α	27/16	$T_1 = (\alpha p\gamma p)$	1.687 5
b					
b	$b = (2,-2,0)$	$(p\gamma p)$	16/9		1.777 778
h					
h	$h = (-2,5,0)$	α	243/128	$\alpha = (Sp^{-1})$	1.898 437
c^1	$\bar{c}^1 = c^1 = (1,0,0)$		2/1		2.0

shows the cumulative build up, while column #6 shows the interval factors/vectors between neighboring lattice tones. Columns #7 and #8 describe the hypothetical 23-tone Carnatic system. Column #7 characterizes the tones, column #8 lists the interval factors for the basis $\alpha = (3, -5, 0)$, $p = (-1, 3, -1)$ and $\gamma = (-2, 1, 2)$. Column #9 lists the standard musical names for the intervals/tones as given in ref. [12].

The hypothetical 23-tone Carnatic scale consists of the prime c and 11 pairs of tones, with each pair of tones related by the Pythagorean vector p. Thus projecting the traditional Carnatic scale along the vector p into the Pythagorean plane 12 tones are obtained. The tone system obtained in this manner corresponds to the 12-tone Pythagorean tone scale, Table 4.1. Table 10.2 contains the following tonal systems:

(a) The traditional 3-dimensional 23-tone Carnatic scale is given by the sequence of intervals

$$\{\alpha, p, \gamma, (\gamma p^{-1})\},$$
$$/\alpha p\gamma p/\alpha p\gamma p/\alpha/p\alpha p(\gamma p^{-1})p/\alpha p\gamma p/\alpha p\gamma p/\alpha/,$$
$$\alpha^7 p^{11}\gamma^4(\gamma p^{-1}) = 2, \qquad N = 23,$$
$$\alpha = (3, -5, 0), \qquad p = (-1, 3, -1), \qquad \gamma = (-2, 1, 2),$$
$$(\alpha p\gamma p) = T_1 = S_1 S_2^2 = s_1^3 s_2^2 = t, \qquad \alpha = Sp^{-1}, \qquad \alpha^7 p^{11}\gamma^4(\gamma p^{-1}) = 2.$$
$$(10.13)$$

Note that if in Eq. (10.13) the sequence of intervals $/p\alpha p(\gamma p^{-1})p/$ is taken as $/p\alpha p\gamma/$ then an $N = 22$-tone Carnatic scale/system is obtained. If the order of the sequence in Eq. (10.13) is changed to

$$/\alpha p\gamma p/\alpha p\gamma p/\alpha/p(\gamma p^{-1})p\alpha p/\alpha p\gamma p/\alpha p\gamma p/\alpha/, \qquad (10.14)$$

then the tone $\bar{f}is$ is obtained in place of the tone $\bar{g}es$. Again, if the sequence $/p(\gamma p^{-1})p\alpha p/$ is replaced by the sequence $/p\gamma\alpha p/$ an $N = 22$ Carnatic scale/system is obtained.

(b) The 12-tone Pythagorean scale: The tones obtained via the sequence Eq. (10.14) consist of the prime c and 11 pairs of tones, with each pair related by the Pythagorean vector p. Thus, if projected into the Pythagorean plane each pair is mapped upon a single tone. The 12 tones thus obtained correspond to the 12-tone Pythagorean scale given in Table 4.1,

$$\{\alpha, (p\gamma p)\}$$
$$/\alpha(p\gamma p)/\alpha(p\gamma p)/\alpha/\alpha(p\gamma p)/\alpha(p\gamma p)/\alpha(p\gamma p)/\alpha/$$
$$\alpha^7(p\gamma p)^5 = 2, \qquad N = 12,$$

$$\alpha = (3, -5, 0) = Sp^{-1} = s_3, \qquad (p\gamma p) = S_1 S_2 p^2 = (-4, 7, 0) = (s_1 s_3),$$
$$\gamma p^{-1} = S_1 S_2 p^{-1} = (-1, -2, 3). \tag{10.15}$$

See also refs. [10], [11].

(c) The 7-tone natural Pythagorean scale: The Pythagorean 7-tone natural diatonic scale is obtained as

$$\{T_1 = t, (Sp^{-1}) = s\},$$
$$/T_1/T_1/(Sp^{-1})/T_1/T_1/T_1/(Sp^{-1})/,$$
$$T_1^5 (Sp^{-1})^2 = 2, \qquad N = 7. \tag{10.16}$$

(d) The 3-tone scale of the ancient Greek Lyre:

$$(T_1 T_1 (Sp^{-1}))/T_1/(T_1 T_1 (Sp^{-1})), \qquad (T_1 T_1 (Sp^{-1}))^2 T_1 = 2.$$

A summary of the results is given in Table 10.2.

The relationship between the three bases $\{\lambda, \mu, \rho\}$, $\{S_1^{-1}, S_1^2 S_2,$ $S_1^{-1} p^{-1}\}$ and $\{\alpha, p, \gamma\}$ for the 116-tone musical system, the 31-tone musical system and the Carnatic musical system, respectively, is given by

$$\begin{array}{llll}
S_1^{-1} = \lambda^5 \mu^3, & S_1^{-1} p^{-1} = \lambda^3 \mu^2, & \lambda = S_1 p^3 = s_1 p^{-1}, \\
S_2 = \lambda^8 \mu^5 \rho, & S_1^2 S_2 = \lambda^{-2} \mu^{-1} \rho, & \mu = S_1^{-2} p^{-5} = s_1^{-2} p^3, \\
S_3 = \lambda^3 \mu^2 \rho, & p = \lambda^2 \mu, & \rho = S_1^2 S_2 p = s_1^3 s_2 p^{-5},
\end{array} \tag{10.17a}$$

$$\begin{array}{ll}
\alpha = Sp^{-1} = \lambda^4 \mu^3 \rho, & T_1 = (\alpha p \gamma p) = S_1 S_2^2 = s_1^3 s_2^2 = t, \ s = s_3, \\
p = S_2/S = \lambda^2 \mu, & T_2 = T_1 p^{-1}, \\
\gamma = S_1 S_2 = \lambda^3 \mu^2 \rho, & S = \alpha p.
\end{array} \tag{10.17b}$$

The tones given by Eqs. (10.1) and (10.2) are obtained, in terms of the basic vectors λ, μ, ρ, Eq. (10.4), by means of the vector equation

$$(n, m, r) = k_1 \lambda + k_2 \mu + k_3 \rho, \qquad k_i, \ i = 1, 2, 3, \text{integers}, \tag{10.18}$$

by substituting for (n, m, r) the desired lattice tone. For the lattice tone $g/c = (0, 1, 0)$ one obtains the set of three equations

$$0 = -6k_1 + 11k_2 - 6k_3,$$
$$1 = 9k_1 - 15k_2 + 4k_3,$$
$$0 = k_1 - 3k_2 + 5k_3. \tag{10.19}$$

These equations yield the values

$$k_1 = 37, \qquad k_2 = 24, \qquad k_3 = 7, \tag{10.20}$$

and thus

$$g/c = (0, 1, 0) = 3/2 = \lambda^{37}\mu^{24}\rho^7, \qquad N = 68. \tag{10.21}$$

Similarly one obtains

$$
\begin{aligned}
a/c &= (0, 0, 1) &= 5/3 &= \lambda^{46}\mu^{30}\rho^9, & N &= 85, \\
e/c &= (-1, 1, 1) &= 5/4 &= \lambda^{20}\mu^{13}\rho^4, & N &= 37, \\
c^1/c &= (1, 0, 0) &= 2/1 &= \lambda^{63}\mu^{41}\rho^{12}, & N &= 116,
\end{aligned}
$$

and

$$f/c = (c^1/c)(g/c)^{-1} = \lambda^{63}\mu^{41}\rho^{12}(\lambda^{37}\mu^{24}\rho^7)^{-1} = \lambda^{26}\mu^{17}\rho^5,$$

$$N = 48. \tag{10.22}$$

By introducing still finer lattice systems it is possible to approximate arbitrarily close any tonal system (see also ref. [13]), like the 12-tone equal-tempered tonal scale ($^{12}\sqrt{2}$) or the 25-tone experimental scale ($^{25}\sqrt{5}$), developed by STOCKHAUSEN, in terms of "natural parameters (intervals)". While the equal tempered scale is based upon the lattice tone $(1, 0, 0) = 2/1$, STOCKHAUSEN's scale is based upon the lattice tone $(1, 1, 1) = (2/1)(3/2)(5/3) = 5$, an octave plus a fifth plus a sixth.

Acknowledgement

The author wishes to thank Professors MIGUEL LORENTE, DIETER FLURY and PETER KRAMER for their interest and their comments.

References

[1] GRUBER, B. J. (2005) Mathematical-Physical Properties of Musical Tone Systems. Sitzungsber. Öst. Akad. Wiss. Wien, math.-nat. Kl., Abt. II **214**: 43–79

[2] RIEMANN, H. (1970) Dictionary of Music, pp. 796–801. Angerer & Co, London, Da Capo Press, New York (Music Theory Reprint Series)

[3] MAZZOLA, G. (1990) Geometrie der Töne. Birkhäuser, Boston, Berlin

[4] LORENTE, M. (1964) Fundamentos físicos de la tonalidad y la consonancia. Rev. Cienc. Aplic. **97**: 97–116; LORENTE, M. (1965) Physical basis of tonality and consonance. V. Int. Congress of Acoustics, Liege; FERNANDEZ-HERRERO, O., LORENTE, M. (2006) Comprobación experimental de la teoria de consonancia y disonancia musical. Rev. Acustica **37**: 5–10

[5] FLURY, D. (1992) Mathematik als Sprache einer Theorie des Hörbewusstseins. In: Klang und Komponist (BIBA, O., SCHUSTER, W., eds.). Hans Schneider, Tutzing

[6] KISHIBE, S. (1984) The Traditional Music of Japan. Ongaku no Tomo-Sha, Tokyo

[7] MALM, W. P. (1967) Music Cultures of the Pacific, the Near East and Asia (History of Music Series). Prentice Hall, New York

[8] HUYGENS-FOKKER, S.: List of Intervals. http://www.xs4all.nl~huygensf/doc/intervals.html

[9] WADE, B. C. (2004) Music in India – the Classical Traditions. Monahor, New Delhi

[10] KRISHNASWAMY, A. (2003) On the Twelve Basic Intervals in South Indian Classical Music. Audio Engineering Society Convention Paper #5903

[11] SAMBAMURTHY, P. (1982) South Indian Music, Vol. 4. The Indian Music Publishing House, Madras

[12] PIERCE, J. R. (1983) The Science of Musical Sound, p. 67 (Scientific American Library). Freeman, New York-San Francisco

[13] See ref. [3], p. 31

Author's address: Prof. Bruno J. Gruber, Visiting Scientist, Institut für Meteorologie und Geophysik, Universität Wien, Hohe Warte 38, 1190 Wien, Austria. *Permanent address*: Emeritus, College of Science, Southern Illinois University, Mailcode 4403, Carbondale, IL 62901, USA. E-Mail: Gruber@siu.edu.

Sitzungsber. Abt. II (2006) 215: 107–125

Sitzungsberichte

Mathematisch-naturwissenschaftliche Klasse Abt. II
Mathematische, Physikalische und Technische Wissenschaften

© Österreichische Akademie der Wissenschaften 2007
Printed in Austria

The Metrics of Prokhorov and Ky Fan for Assessing Uncertainty in Inverse Problems

By

Andreas Hofinger

(Vorgelegt in der Sitzung der math.-nat. Klasse am 12. Oktober 2006
durch das w. M. Heinz Engl)

Abstract

To assess the quality of solutions in stochastic inverse problems, a proper measure for the distance of random variables is essential.

The aim of this note is the comparison of the metrics of KY FAN and PROKHOROV with other concepts such as expected values, probability estimates and almost sure convergence.

In ill-posed problems one aims to find an appropriate solution $x^{\dagger}$ to an equation of the form

$$F(x) = y,$$

when the operator F is not continuously invertible. Therefore, the problems of interest are unstable; when only noisy data y^{δ} are available, special techniques (so-called regularization methods) must be applied to obtain regularized solutions x_{α}^{δ} that are reasonable approximations to $x^{\dagger}$. To assess the quality of different regularization methods, in the theory of ill-posed problems convergence rate results, i.e., results of the form

$$\|x^{\dagger} - x_{\alpha}^{\delta}\| = \mathcal{O}(f(\|y - y^{\delta}\|)),$$

are an accepted quality criterion (see [12] for an introduction into this topic). So in a nutshell, in the deterministic theory of inverse problems the aim is to bound the distance between desired and regularized solution, in terms of the distance between exact and noisy data.

When the deterministic theory of inverse problems is to be extended to a stochastic setup, a question of utmost importance is how to measure distances of random variables, since now $x^\dagger$, x_α^δ, y, $\ldots$ are replaced by their stochastic counterparts $x^\dagger(\omega)$, $x_\alpha^\delta(\omega)$, $y(\omega)$, $\ldots$. In [18] an approach was presented that performs this extension using the metrics of KY FAN and PROKHOROV. In the following we collect some general results from [18] about these metrics.

The first section of this note introduces the metrics of KY FAN and PROKHOROV and describes general relations and differences between these metrics. In a second section we briefly compare the Ky Fan metric with other qualitative and quantitative concepts for measuring convergence. The final section gives a detailed quantitative comparison of convergence in expectation and in the Ky Fan metric.

1. The Metrics of Prokhorov and Ky Fan

Let us first introduce the metrics of PROKHOROV and KY FAN. As we will see later, these two are closely related, but while the latter works with random variables, the Prokhorov metric is concerned only with the underlying distributions.

1.1. The Prokhorov Metric

This metric does not directly work on the space of random variables, but on the underlying induced *distributions*. Suppose we are given two random variables $x(\omega)$ and $\tilde{x}(\omega) \in X$ for $\omega \in \Omega$ and probability space $(\Omega, \mathcal{A}, \mu)$. Via the measure μ on Ω we can define two corresponding measures μ_x and $\mu_{\tilde{x}}$ (the so-called *distributions* of x and $\tilde{x}$) on the space X: For a Borel set $B \subset X$ we define

$$\mu_x(B) := \mu(x^{-1}(B)) := \mu\{\omega \in \Omega \mid x(\omega) \in B\}$$

and

$$\mu_{\tilde{x}}(B) := \mu(\tilde{x}^{-1}(B)) := \mu\{\omega \in \Omega \mid \tilde{x}(\omega) \in B\}.$$

Instead of measuring the pointwise distance of $x(\omega)$ and $\tilde{x}(\omega)$ directly, we can use this lifting onto spaces of probability measures and compute the distance of the respective measures there, using an appropriate metric.

Distances between probability measures can be defined in numerous ways, see e.g. [17] for an overview. In [18] the aim was to develop a theory for stochastic inverse problems, that contains the deterministic one as a special case. For this sake a concept is needed that is applicable to point-measures, because such point-measures correspond to "constant random variables", which are essentially deterministic quantities. As the following remarks show, it is therefore important that the chosen metric metrizes the weak-star topology.

Remark 1.1. Consider an interval $I \subset \mathbb{R}$ and suppose that a sequence $(x_k)_{k=1}^{\infty}$, $x_k \in I$ converges to some $x \in I$. For which topology do the corresponding point-measures δ_{x_k} converge?

Every probability measure μ defines a continuous linear functional on the space of continuous functions $C(I)$ via

$$\mu(f) := \int_I f(x)\,d\mu(x),$$

therefore every measure is an element of the dual space $C(I)^*$ of $C(I)$. Since $C(I)$ is not reflexive (cf. [28, 30]) there are at least 3 different possibilities for defining convergence of measures in $C(I)^*$: norm-, weak-, and weak-star-convergence. Let in the following δ_{x_k} and δ_x denote point-measures, associated with the points in the sequence $(x_k) \subset I \setminus \{x\}$ and their limit x.

• *Norm topology*: A norm on $C(I)^*$ can be defined via elements of the pre-dual space $C(I)$ as

$$\|\mu_1 - \mu_2\| = \sup_{f \in C(I)} \frac{|\int f(x)\,d(\mu_1 - \mu_2)|}{\|f\|_{C(I)}}.$$

For given k we can always find a continuous function $f_k(\cdot)$, $\|f_k\| \le 1$, with $f_k(x) = 1$ and $f_k(x_k) = 0$ (e.g. a piecewise linear interpolant). Since $\int f_k(x)\,d(\delta_x - \delta_{x_k}) = f_k(x) - f_k(x_k)$, the distance between δ_{x_k} and δ_x in the norm remains constant and equal to 1; the point-measures δ_{x_k} do not converge to δ_x in the norm topology.

• *Weak topology*:[1] To investigate weak convergence we consider convergence of $\phi(\delta_{x_k})$ to $\phi(\delta_x)$ where $\phi \in C(I)^{**}$, i.e., ϕ is an element of the dual of $C(I)^*$. One particular functional on $C(I)^*$ is defined by the point evaluation

$$\phi_x \colon C(I)^* \to \mathbb{R}$$

$$\mu \mapsto \phi_x(\mu) := \mu(\{x\}).$$

This functional is linear and has norm 1, as can be seen via the Riesz representation theorem ([28, 30]). With this choice we obtain for all

---

[1] Note that the common notion of *weak topology* for probability distributions typically means the weak-star topology discussed below; we use the functional analytic terminology.

$k \in \mathbb{N}$ that $\phi_x(\delta_{x_k}) = 0$, while at the same time $\phi_x(\delta_x) = 1$, thus δ_{x_k} does not converge weakly to δ_x.

• *Weak-star topology*: Here we measure convergence of δ_{x_k} to δ_x by applying these measures to continuous functions f. For any $f \in C(I) \subsetneq C(I)^{**}$ we have

$$|\delta_{x_k}(f) - \delta_x(f)| := \left| \int f(x)\, d(\delta_{x_k} - \delta_x) \right| = |f(x_k) - f(x)|.$$

Since f is continuous we obtain that $\delta_{x_k} \to \delta_x$, whenever $x_k \to x$, so the point-measures δ_{x_k} do converge to δ_x in the weak-star topology.

Thus, the weak-star topology is weak enough such that $\delta_{x_k} \to \delta_x$ whenever $x_k \to x$.

A complementary question is considered in the following remark.

Remark 1.2. Given a probability space $(\Omega, \mathcal{A}, \mu)$, we denote the space of probability measures μ_x, which are actually distributions of random variables x, $x \colon \Omega \to X$, by $\mathcal{M}_{(\Omega, \mathcal{A}, \mu)}(X)$. This space is not necessarily equal to the space of all probability measures on X (denoted by $\mathcal{M}(X)$), since not every probability measure needs to be the distribution of a random variable on $(\Omega, \mathcal{A}, \mu)$.

But what can be said about a probability measure that is, in some sense, the limit of distributions of random variables, will it also be the distribution of a random variable? I.e., if a sequence μ_{x_n} converges to some $\tilde{\mu} \in \mathcal{M}(X)$, does there exist a random variable $\tilde{x}$, $\tilde{x} \colon \Omega \to X$, such that $\tilde{\mu} = \mu_{\tilde{x}}$? For the case of convergence in the weak-star topology, this question is answered affirmatively in [14].

Thus, the weak-star topology is strong enough such that $\mathcal{M}_{(\Omega, \mathcal{A}, \mu)}(X)$ is a sequentially closed subset of $\mathcal{M}(X)$.

Although many different distances on probability spaces are available, only few of them metrize the weak-star topology. Among them are the Prokhorov metric and the Wasserstein or bounded Lipschitz metric. The result of STRASSEN (see Theorem 1.6) gives connections between the distance of the measures on the probability space measured in the Prokhorov metric, and the distance of corresponding random variables measured in the Ky Fan metric, and is the reason why we finally pick this metric.

Definition 1.3 (Prokhorov metric). The distance of two measures μ_1, μ_2 in the Prokhorov metric is defined as (see, e.g., [5, 9, 19, 22])

$$\rho_P(\mu_1, \mu_2) := \inf\{\varepsilon > 0 \mid \mu_1(B) \leq \mu_2(B^\varepsilon) + \varepsilon, \forall \text{ Borel sets } B \subset \Omega\}.$$

$$(1)$$

Here $B^\varepsilon = \{x \,|\, d(x, B) < \varepsilon\}$, where $d(x, B)$ is the distance of x to B, i.e., $d(x, B) = \inf_{z \in B} \|x - z\|$.

The use of B^ε in the definition above is essential: The Prokhorov distance of two measures μ_1 and μ_2 is small, when the probability of similar events (B^ε) is similar up to a small quantity ("$+\varepsilon$"). In contrast, the *total variation distance* (cf. [17]) measures if the probability of *the same* event (B) is similar ("$+\varepsilon$"). Consequently, when x_k, $x_k \neq x$, converges to x the corresponding point-measures δ_{x_k} converge to δ_x in the Prokhorov metric, but they do not in the total variation distance.

Having defined a metric on the space of probability measures, an interesting question is the following: Consider an operator $F \colon X \to Y$, and a sequence of random variables (x_k), with $x_k \to x$, where convergence is measured in the Prokhorov metric. Under which continuity assumptions on F does the random variable $F(x_k)$ converge to $F(x)$ (again in the Prokhorov metric), and can this convergence be quantified? This question is answered in [15, 29]. As demonstrated in [13, 15] the answer to this question is also relevant for regularization theory of stochastic inverse problems, because it can be used to derive convergence rate results.

1.2. The Ky Fan Metric

While the Prokhorov metric works with distributions, the Ky Fan metric uses random variables to define distances. This metric is defined as follows.

Definition 1.4 (Ky Fan metric). The distance of two random variables ξ_1, ξ_2 in the Ky Fan metric is defined as ([16], also [9])

$$\rho_K(\xi_1, \xi_2) := \inf\{\varepsilon > 0 \,|\, \mu\{\omega \in \Omega \,|\, d(\xi_1(\omega), \xi_2(\omega)) > \varepsilon\} < \varepsilon\}. \quad (2)$$

Convergence in the Ky Fan metric is a quantitative version of *convergence in probability* (see below).

Let us first give a short interpretation of the Ky Fan distance. If ξ_1 and ξ_2 have distance $\rho_K(\xi_1, \xi_2) \leq \varepsilon$, this implies that

- with high probability (namely $1 - \varepsilon$) the realizations of the random variables have distance $d(\xi_1(\omega), \xi_2(\omega)) \leq \varepsilon$,
- with low probability (at most ε), the distance between $\xi_1(\omega)$ and $\xi_2(\omega)$ may be larger than ε.

In particular, the second point is of interest: The Ky Fan distance between ξ_1 and ξ_2 may be small, although on a set of positive probability the distance of $\xi_1(\omega)$ and $\xi_2(\omega)$ might be arbitrarily large.

In contrast, the expected value is influenced by all events that have positive probability. In particular for non-normally distributed random variables this can make a significant difference. For instance in Section 3 we construct an example for which $\rho_K(\xi_1, \xi_2) \to 0$ while $\mathbb{E}(\|\xi_1 - \xi_2\|^2)$ remains constant or even tends to infinity. This is also relevant for convergence rates for stochastic inverse problems, as a numerical example in [18, Ch. 5] shows:

Example. The quality of the solutions there is measured by some parameter s, where $s = 0$ corresponds to a deterministic problem with well-behaved solution; with growing s the probability of "bad" solutions increases. While now the Ky Fan metric gives

$$\rho_K\left(x^\delta_{k_*(\delta)}, x^\dagger\right)_{H^1(I)} = \mathcal{O}\left(\delta^{2\nu/(2\nu+1+s)}\right),$$

i.e., a convergence rate that slows down gradually for increasing s, the rate in expectation is given as

$$\mathbb{E}\left(\|x^\dagger - x^\delta_{k_*}\|^2_{H^1(I)}\right)^{1/2} \le \begin{cases} \mathcal{O}\left(\delta^{2\nu/(2\nu+1)}\right), & 0 < s < \nu + \tfrac{1}{2}, \\ \infty, & \nu + \tfrac{1}{2} \le s, \end{cases}$$

i.e., it remains constant for a while, and switches to non-convergence suddenly.[2]

So on the one hand measuring convergence in expectation, one is restricted to a smaller class of random variables. On the other hand, the expected value does not give information about particular realizations, while the Ky Fan distance gives a good bound with probability $1 - \varepsilon$.

1.3. Relations Between Prokhorov and Ky Fan Metric

Let us in the following consider some connections and differences between the Ky Fan metric and the Prokhorov metric.

In case we are interested in the distance of a random variable to a point, or respectively, the distance of a distribution to a point-measure, it turns out that the Ky Fan distance and the Prokhorov distance are equal.

---

[2] Of course the involved coefficients in the $\mathcal{O}(\cdot)$-notation do not remain constant. But since these constants are in practice unavailable, the focus in inverse-problems theory is on the appearing exponents in the convergence rate (see [12]).

Proposition 1.5. *Let ξ_1, ξ_2 be two random variables with distributions μ_1, μ_2. Let one of the random variables be constant. Then*

$$\rho_K(\xi_1, \xi_2) = \rho_P(\mu_1, \mu_2).$$

Proof. It suffices to show that $\rho_K(\xi_1, \xi_2) \leq \rho_P(\mu_1, \mu_2)$ (see Proposition 1.7 below for the converse estimate).

Suppose that ξ_1 is constant, i.e., $\xi_1(\omega) = x_1$ for almost all $\omega \in \Omega$. According to Definition 1.3 we have for arbitrary Borel-sets B and $\varepsilon > \rho_P(\mu_1, \mu_2)$

$$\mu_1(B) < \mu_2(B^\varepsilon) + \varepsilon.$$

In particular this also holds for the set $B = \{x_1\}$. For this choice we obtain

$$
\begin{aligned}
1 - \varepsilon < \mu_2(B^\varepsilon) \\
&= \mu\{\omega \in \Omega \mid \xi_2(\omega) \in B^\varepsilon\} \\
&= \mu\{\omega \in \Omega \mid d(\xi_2(\omega), x_1) \leq \varepsilon\}.
\end{aligned}
$$

Via Definition 1.4 this implies $\rho_K(\xi_1, \xi_2) \leq \varepsilon$. Taking the infimum with respect to $\varepsilon > \rho_P(\mu_1, \mu_2)$ concludes the proof. $\square$

If we are interested in the distance of two genuine random variables, more effort is necessary to connect the Prokhorov metric and the Ky Fan metric. The following theorem, originally obtained by STRASSEN [27] and extended by DUDLEY [8], is an important tool for this task. (The proof of the following result can also be found in [9, ch. 11.6].)

Theorem 1.6 (STRASSEN). *Let $(\mathcal{X}, d)$ be a separable metric space and $\mathcal{M}(\mathcal{X})$ be the set of Borel probability measures on $\mathcal{X}$, μ_1, μ_2 be elements of this space. Let $\alpha \geq 0$ and $\beta \geq 0$. Then the following statements are equivalent:*

(i) $\mu_1(B) \leq \mu_2(B^{\alpha]}) + \beta$ for all Borel sets B.

(ii) For any $\varepsilon > 0$ there exists $\mu_{1/2} \in \mathcal{M}(\mathcal{X} \times \mathcal{X})$ with $\mu_1 = \mu_{1/2} \circ \pi_1^{-1}$, $\mu_2 = \mu_{1/2} \circ \pi_2^{-1}$ and $\mu_{1/2}(d(x_1, x_2) > \alpha + \varepsilon) \leq \beta + \varepsilon$. Here π_1, π_2 are the natural projections from $\mathcal{X} \times \mathcal{X}$ onto $\mathcal{X}$, i.e., $\pi_1(x_1, x_2) = x_1$, $\pi_2(x_1, x_2) = x_2$.

Here $B^{\alpha]}$ denotes the closure of B^α. If furthermore $\mathcal{X}$ is complete, this equivalence also holds for the case $\varepsilon = 0$.

This theorem immediately gives the following relation between the Prokhorov metric and the Ky Fan metric [9, 25, 27].

Proposition 1.7. *Let the assumptions of Theorem 1.6 be satisfied, and $\mathcal{X}$ be complete. Then with the notation of Theorem 1.6 the following statements are equivalent:*

(i) The Prokhorov distance of two measures μ_1, μ_2 satisfies $\rho_P(\mu_1, \mu_2) \leq \varepsilon$.

(ii) There exist random variables ξ_1 and ξ_2 such that μ_1 and μ_2 are the distributions of ξ_1 and ξ_2, respectively, and $\rho_K(\xi_1, \xi_2) \leq \varepsilon$.

Thus, we always have $\rho_P(\mu_1, \mu_2) \leq \rho_K(\xi_1, \xi_2)$; for given μ_1, μ_2 we can find ξ_1, ξ_2 and a joint distribution $\mu_{1/2} \in \mathcal{M}(\mathcal{X} \times \mathcal{X})$ such that equality holds.

Proof. Let $\rho_P(\mu_1, \mu_2) = \varepsilon$. Then by Definition 1.3 (note that the set on the right is closed)

$$\mu_1(B) \leq \mu_2(B^{\varepsilon]}) + \varepsilon.$$

The Theorem of STRASSEN now guarantees existence of $\mu_{1/2} \in \mathcal{M}(\mathcal{X} \times \mathcal{X})$ with the property

$$\mu_{1/2}(d(x_1, x_2) > \varepsilon) \leq \varepsilon,$$

so $\rho_K(\pi_1, \pi_2) \leq \rho_P(\mu_1, \mu_2)$, where the projections π_1 and π_2 are seen as random variables from $\mathcal{X} \times \mathcal{X}$ to $\mathcal{X}$. The distributions of π_1 and π_2 are μ_1 and μ_2, respectively.

For the converse implication, suppose $\rho_K(\xi_1, \xi_2) = \varepsilon$, and choose $\delta > 0$. Via STRASSEN's theorem this implies

$$\mu_1(B) \leq \mu_2(B^{\varepsilon + \delta]}) + \varepsilon + \delta.$$

To obtain an open set on the right-hand side, we can estimate

$$\mu_1(B) \leq \mu_2(B^{\varepsilon + 2\delta}) + \varepsilon + 2\delta.$$

This holds for all Borel-sets B, and therefore $\rho_P(\mu_1, \mu_2) \leq \rho_K(\xi_1, \xi_2) + 2\delta$. The estimate follows by building the infimum with respect to $\delta > 0$. $\qquad\square$

Since the Prokhorov distance of two measures μ_1 and μ_2 is so closely related to the Ky Fan distance of corresponding random variables ξ_1 and ξ_2, the appearance of many convergence results remains unchanged when the Prokhorov metric is replaced by the Ky Fan metric. For instance the structure of the lifting result presented in [13, Thm. 2.1] remains unchanged when $\rho_P(\cdot, \cdot)$ is replaced by $\rho_K(\cdot, \cdot)$ (see [18, Thm. 1.15]).

Let us now turn to *differences* between the Prokhorov metric and the Ky Fan metric.

As already noted, the Ky Fan distance gives information about concrete realizations of random variables, while the Prokhorov distance only works with the underlying distributions. To underline this difference, consider tossing a fair coin, where we denote the outcome "heads" by $+1$ and "tails" by -1. Now suppose that (maybe due to an assembly error) our measurement device always observes the opposite outcome, i.e., it measures "heads" when there is "tails" and vice versa. What is the measurement error for this experiment?

- For the Ky Fan metric the error clearly is equal to 1: In 100% of the cases we measure $+1$ instead of -1 and vice versa. So with probability 1 the measurement error is larger than 1 (cf. Definition 1.4).
- In contrast, for the Prokhorov metric the error is 0, i.e., the observation seems to be noise free (!): In 50% of the cases we observe the outcome $+1$, in 50% of the cases we observe the outcome -1. Although each single measurement is wrong, in total we find that the experiment describes a binomial distribution with parameter $\frac{1}{2}$, which is the correct observation.

This example indicates that results in the Ky Fan metric will typically be suitable when the focus is on random variables. In particular, an estimate in the Ky Fan metric gives a *confidence interval* for concrete realizations; if $\rho_K(\xi_1, \xi_2) \le \varepsilon$, with probability $1 - \varepsilon$ the realizations $\xi_1(\omega)$ and $\xi_2(\omega)$ have distance less than ε.

In contrast, convergence results in the Prokhorov metric are of interest, when the observed data are a distribution. Consider a biological system where the growth behavior of cells is to be analyzed (cf. [1]). While it is possible to determine, e.g., the distribution of the cell sizes within the system (maybe even in a time-dependent way), there may be no means to track the behavior of only a single cell. Clearly, it is not possible to use this input data to determine parameters of individual cells; so one cannot ask for more than convergence of the reconstructed distribution to the correct value. The Prokhorov metric is the right tool to measure convergence in such a situation.

1.4. Further Properties of the Prokhorov Metric

In this section we mention some additional properties of the Prokhorov metric. Suppose in the following that $\mathcal{X}$ is a vector space equipped with a metric, and $\mathcal{M}(\mathcal{X})$ is the set of probability measures on this space. Some strong relations hold between the space $\mathcal{X}$, and

$\mathcal{M}(\mathcal{X})$ equipped with the metric $\rho_P(\cdot, \cdot)$; these results were obtained by PROKHOROV in [23] (cf. [22]).

- $\mathcal{M}(\mathcal{X})$ can be metrized as a separable metric space if and only if $\mathcal{X}$ is itself a separable metric space.
- $\mathcal{M}(\mathcal{X})$ is a Polish space (i.e., separable, metric and *complete*) if and only if $\mathcal{X}$ is Polish.
- $\mathcal{M}(\mathcal{X})$ is a *compact* metric space if and only if $\mathcal{X}$ is compact metric.

The last fact was used in [1] to show that a certain approximation scheme leads to convergence in the Prokhorov metric: In the considered case an ODE in a biological system is influenced by some real valued random variable $C(\omega)$, each individual carrying a different realization of C. The goal in [1] was now to recover the distribution of $C(\omega)$ via a time-dependent measurement of the evolution of the system. Since in [1] $C(\omega)$ was confined to some closed interval, we obtain from the above statement that also the corresponding distributions come from a compact space, and thus, have convergent subsequences in the Prokhorov metric. If the desired quantity was no real or (finite dimensional) vector-valued random variable C, but a random function this approach would fail, since closed bounded sets in infinite dimensional spaces need not be compact.

2. Comparison with Other Concepts

In the following we consider various other approaches that are common when describing convergence of random variables. This comparison is split into qualitative and quantitative concepts. As qualitative ones we choose the two common concepts *convergence almost surely* and *convergence in probability*; as quantitative ones we choose such concepts that have been used in the theory of stochastic inverse problems (see, e.g., [6, 26]).

2.1. Qualitative Concepts

Convergence almost surely: A sequence x_k converges to x almost surely (a.s.), when for almost all ω

$$\|x_k(\omega) - x(\omega)\| \to 0, \tag{3}$$

i.e., except on a null set, $x_k(\omega)$ converges point-wise to $x(\omega)$. Almost sure convergence implies convergence in the Ky Fan metric; the converse is true for subsequences (Propositions 2.1 and 2.2 below).

Convergence in probability: A sequence x_k converges to x in probability, when for all $\varepsilon > 0$

$$\mu\{\omega \in \Omega \mid \|x_k(\omega) - x(\omega)\| > \varepsilon\} \to 0.$$

Comparing this definition with (2), it can be seen easily that this type of convergence is equivalent to convergence in the Ky Fan metric (see also [16]).

Relations between almost sure convergence and convergence in probability are well-known (see e.g., [2, 3, 7, 9, 11]). Because convergence in the Ky Fan metric is a quantitative version of convergence in probability analogous results hold for the Ky Fan metric as well. In the following we discuss these relations between almost sure convergence and convergence in the Ky Fan metric. The first theorem follows immediately from the analogous result for convergence in probability.

Proposition 2.1. *Let $x_k \to x$ almost surely. Then $\rho_K(x_k, x) \to 0$.*

The converse result is not true. Nevertheless, it is well known that convergence in probability implies almost sure convergence at least of *subsequences* (see, e.g., [9, 11]). The next proposition gives a quantitative version of this statement for the Ky Fan metric[3].

Proposition 2.2. *Let x_k converge to x in the Ky Fan metric. Then for any $\eta > 0$ and $\varepsilon > 0$ there exist Ω_ε, $\mu(\Omega_\varepsilon) \geq 1 - \varepsilon$, and a subsequence x_{k_j} with*

$$\|x_{k_j}(\omega) - x(\omega)\| \leq (1 + \eta)\rho_K(x_{k_j}, x) \qquad \text{for all} \quad \omega \in \Omega_\varepsilon.$$

Furthermore there exists a subsequence that converges to x almost surely.

Proof. Set $\delta_k := (1 + \eta)\rho_K(x_k, x)$. By Definition 1.4, for given δ_k, there exists a set Ω_{δ_k} with

$$\mu(\Omega_{\delta_k}) \geq 1 - \delta_k, \qquad \text{and} \qquad \omega \in \Omega_{\delta_k} \Longrightarrow \|x(\omega) - x_k(\omega)\| \leq \delta_k.$$

In general we cannot deduce convergence of $x_k(\omega)$ to $x(\omega)$ for $\omega \in \Omega_{\delta_k}$ and $\delta_k \to 0$, since the sets Ω_{δ_k} may have empty intersection for $\delta_k \to 0$. Thus we need the following construction. For arbitrary $\varepsilon > 0$ and $\delta_k \to 0$ we pick a subsequence (δ_{k^j}) with $\sum_{j=1}^\infty \delta_{k^j} \leq \varepsilon$, and introduce the set

$$\Omega_\varepsilon := \bigcap_{j=1}^{\infty} \Omega_{\delta_{k^j}},$$

---

[3] Presumably, the following result has not been explicitly derived before, since only little attention was paid to the Ky Fan metric in the past.

which is a subset of every $\Omega_{\delta_{kj}}$. This set has measure $\mu(\Omega_\varepsilon) \geq 1 - \varepsilon$ since

$$\mu\left(\bigcap_{j=1}^{\infty}\Omega_{\delta_{kj}}\right) = \mu\left(\Omega\backslash\bigcup_{j=1}^{\infty}\left(\Omega\backslash\Omega_{\delta_{kj}}\right)\right) = 1 - \mu\left(\bigcup_{j=1}^{\infty}\left(\Omega\backslash\Omega_{\delta_{kj}}\right)\right)$$

$$\geq 1 - \sum_{j=1}^{\infty}\mu\left(\Omega\backslash\Omega_{\delta_{kj}}\right) \geq 1 - \varepsilon.$$

Since Ω_ε is a subset of every $\Omega_{\delta_{kj}}$ we have

$$\forall \omega \in \Omega_\varepsilon \subseteq \Omega_{\delta_{kj}}: \quad \|x(\omega) - x_{k_j}(\omega)\| \leq \delta_{kj},$$

which proves the first statement.

For the second statement, consider the set N on which x_{k_j} does not converge to x. This set is given as

$$N = \{\omega \mid \exists \tilde{\varepsilon} > 0 \ \forall j_0 \in \mathbb{N} \ \exists j \geq j_0 : \|x_{k_j}(\omega) - x(\omega)\| \geq \tilde{\varepsilon}\}$$

$$= 1 - \{\omega \mid \forall \tilde{\varepsilon} > 0 \ \exists j_0 \in \mathbb{N} \ \forall j \geq j_0 : \|x_{k_j}(\omega) - x(\omega)\| < \tilde{\varepsilon}\}.$$

Similarly to Ω_ε above, we define the set Ω_{j_0} as

$$\Omega_{j_0} := \bigcap_{j \geq j_0}\Omega_{\delta_{kj}} = \{\omega \mid \forall j \geq j_0 : \|x_{k_j}(\omega) - x(\omega)\| < \delta_{kj}\}$$

$$\subseteq \{\omega \mid \forall \tilde{\varepsilon} > 0 \ \exists j_0 \in \mathbb{N} \ \forall j \geq j_0 : \|x_{k_j}(\omega) - x(\omega)\| < \tilde{\varepsilon}\}.$$

Since the sequence x_{k_j} is uniformly convergent to x on the set Ω_{j_0}, we can estimate

$$\mu(N) = 1 - \mu(\{\omega \mid \forall \tilde{\varepsilon} > 0 \ \exists j_0 \in \mathbb{N} \ \forall j \geq j_0 : \|x_{k_j}(\omega) - x(\omega)\| < \tilde{\varepsilon}\})$$

$$\leq 1 - \mu(\{\omega \mid \forall j \geq j_0 : \|x_{k_j}(\omega) - x(\omega)\| < \delta_{kj}\})$$

$$= 1 - \mu\left(\bigcap_{j \geq j_0}\Omega_{\delta_{kj}}\right)$$

$$= \mu\left(\bigcup_{j \geq j_0}\Omega\backslash\Omega_{\delta_{kj}}\right) \leq \sum_{j \geq j_0}\delta_{kj}.$$

Since $\sum \delta_{kj} \leq \varepsilon$, and this sum is absolutely convergent, we obtain

$$\sum_{j \geq j_0}\delta_{kj} \to 0 \quad \text{as} \quad j_0 \to \infty.$$

Hence N is a null set; x_{k_j} converges to x almost surely. $\qquad\square$

2.2. Quantitative Concepts

Now we turn to quantitative concepts to measure convergence of random variables. Two common concepts are expected values and probability estimates.

Convergence in expectation: Here we look for bounds on the expected value of the distance, defined as

$$\mathbb{E}(\|x_k - x\|^2) := \int_\Omega \|x_k(\omega) - x(\omega)\|^2 \, d\mu(\omega). \tag{4}$$

For the case $X = L_2(I)$, we find that $\mathbb{E}(\|\cdot\|^2)$ is a (weighted) norm on the product space $L_2(\Omega \times I)$.

Probability estimates: Similarly as for the Ky Fan metric, the space Ω is split into parts with high and low probability, the resulting estimates have the form[4]

$$\mathbb{P}(\|x_k - x\| \le \varepsilon_1(p)) > 1 - \varepsilon_2(p). \tag{5}$$

Here $\varepsilon_1(p)$ and $\varepsilon_2(p)$ are functions of one or more parameters p. Typically, these functions are continuous.

The concept of expectation can be a too restrictive notion of convergence. A first indication is the fact that almost sure convergence as in (3) does not guarantee that also $\mathbb{E}(\|x_k - x\|^2)$ tends to 0; even worse, we could have that $\mathbb{E}(\|x_k - x\|^2) = \infty$ and remains unbounded, no matter how large k is chosen (cf. the examples in Section 3).

Therefore the second error measure can be considered more natural. Assuming that x_k converges to x almost surely, we have for any fixed ε_1 that $\mathbb{P}(\|x_k - x\| \le \varepsilon_1)$ tends to 1. Vice versa, for fixed ε_2 and arbitrary small $\varepsilon_1 > 0$ we can find $k \in \mathbb{N}$ with $\mathbb{P}(\|x_k - x\| \le \varepsilon_1) > 1 - \varepsilon_2$. A drawback of this concept is that *two* parameters are necessary to describe the distance; estimates of the form (5) do not form a metric.

But fortunately, via Definition 1.4 these can be translated into estimates in the Ky Fan metric, and via Proposition 1.7 also into the Prokhorov metric. To do this, one has to solve the equation

$$\varepsilon_1(p) = \varepsilon_2(p)$$

---

[4] Depending on the context we will sometimes use the notion of probabilities instead of measures, but of course these are only two different words for the same meaning and $\mathbb{P}(B) \equiv \mu(B)$.

for p. The resulting solution gives some $\varepsilon(p) = \varepsilon_1(p) = \varepsilon_2(p)$ with $\rho_K(x_k, x) \leq \varepsilon(p)$. The task of solving this equation is often non-trivial, but as its outcome we obtain a distance between x_k and x in terms of a metric, which allows us to investigate the speed of convergence.

Thus, concluding we will find that the Ky Fan metric gives:

- A concept that is better suited for treating stochastic problems than the expected value, especially when the distributions of the appearing variables may have 'fat tails', i.e., when there is a higher probability for large values to occur than for a normally distributed random variable. In particular, the Ky Fan distance is always finite, whereas the expected value (4) can be unbounded; pointwise convergence implies convergence in the Ky Fan metric (cf. e.g. the constructions in the next section).
- A framework that translates estimates as in (5) to an interpretable setup. This allows comparison of new stochastic results with the classical deterministic ones.
- A setting that contains the deterministic results as a special case.

Let us in the following make a more detailed comparison between the Ky Fan metric and the expected value.

3. Expectation Versus Ky Fan Metric

3.1. Non-Convergence of Expectation

In the following we construct a sequence x_k that converges to 0 in the Ky Fan metric, but does not converge in expectation. For this construction we define the random variable $x(\omega)$ as

$$x(\omega) = \frac{C(\alpha)}{\omega^\alpha}, \tag{6}$$

where ω is uniformly distributed on $\Omega = [0, 1]$. Observe that $x(\omega)$ is unbounded on the interval $[0, 1]$ for $\alpha > 0$; nevertheless, clearly for $C(\alpha) \to 0$ also $x(\omega)$ tends to 0 on $(0, 1]$ point-wise.

We now compute the expected value of the distance of $x(\omega)$ to 0. In the first case we look for the expectation of the absolute value of the error and obtain

$$\mathbb{E}(|x - 0|) = \int_\Omega \frac{C(\alpha)}{\omega^\alpha} \, d\omega = C(\alpha) \int_\Omega \omega^{-\alpha} \, d\omega$$
$$= \begin{cases} C(\alpha)/(1 - \alpha), & \alpha < 1, \\ \infty, & \alpha \geq 1. \end{cases}$$

For the second case we investigate the square of the absolute value

$$\mathbb{E}(|x - 0|^2) = \int_\Omega \frac{C(\alpha)^2}{\omega^{2\alpha}} \, d\omega = C(\alpha)^2 \int_\Omega \omega^{-2\alpha} \, d\omega$$

$$= \begin{cases} C(\alpha)^2/(1 - 2\alpha), & \alpha < \frac{1}{2}, \\ \infty, & \alpha \geq \frac{1}{2}. \end{cases}$$

Observe that, for α sufficiently large, in both cases the expected value may be infinite, although $x(\omega)$ tends to 0 pointwise.

Now we compute the Ky Fan distance, therefore we have to find the smallest t such that

$$\mathbb{P}(|x(\omega) - 0| > t) < t.$$

Inserting (6) we have for the term on the left

$$\mathbb{P}\left(\frac{C(\alpha)}{\omega^\alpha} > t\right) = \mathbb{P}\left(\frac{1}{\omega} > \left(\frac{t}{C(\alpha)}\right)^{1/\alpha}\right) = \left(\frac{C(\alpha)}{t}\right)^{1/\alpha},$$

which further leads to the relation

$$\left(\frac{C(\alpha)}{t}\right)^{1/\alpha} = t.$$

Solving this equation for t we obtain for the Ky Fan distance the expression

$$\rho_K(x, 0) = C(\alpha)^{1/(1+\alpha)}.$$

Now define a sequence of random variables $x_k(\omega) = C(\alpha_k)/\omega^{\alpha_k}$ with some $\alpha_k \nearrow 1$ and $C(\alpha_k) := 1 - \alpha_k$. Then we obtain that in the Ky Fan metric this sequence tends to 0, indeed

$$\rho_K(x_k, 0) = (1 - \alpha_k)^{1/(1+\alpha_k)} \leq (1 - \alpha_k)^{1/2} \to 0. \tag{7}$$

So the random variable x_k converges to 0 in the Ky Fan metric, nevertheless the expectation does not tend to 0, it even remains constant

$$\mathbb{E}(|x_k - 0|) = \frac{C(\alpha_k)}{1 - \alpha_k} = 1 \nrightarrow 0. \tag{8}$$

The choice $\alpha_k \nearrow \frac{1}{2}$ and $C(\alpha_k) := 1 - 2\alpha_k$ yields the analogous result for $\mathbb{E}(|x_k - 0|^2)$. Observe that for the latter choice $\mathbb{E}(|x_k - 0|^2)$ remains constant while $\mathbb{E}(|x_k - 0|)$ tends to 0. Furthermore observe that $C(\alpha_k) := (1 - \alpha_k)^{1/2}$ even leads to divergence in (8), while maintaining convergence in (7).

Clearly, the presented results arise from the unboundedness of the involved random variables. If the distance of the random variables

ξ_1 and ξ_2 is bounded a-priori, we obtain from (2) (with arbitrary $\varepsilon > \rho_K(\xi_1, \xi_2)$)

$$\mathbb{E}(\|\xi_1 - \xi_2\|^2) := \int_\Omega \|\xi_1(\omega) - \xi_2(\omega)\|^2 \, d\mu(\omega)$$
$$\leq (1 - \varepsilon)\varepsilon^2 + \varepsilon \sup_{\omega \in \Omega} \|\xi_1(\omega) - \xi_2(\omega)\|^2 \leq C\varepsilon.$$

So for bounded random variables, convergence in the Ky Fan metric implies convergence of the expected value.

3.2. Convergence via Markov's Inequality

We have seen that a sequence may converge in the Ky Fan metric, without the need to converge in expectation. On the other hand, comparing the representation (4) with (2) it seems clear that whenever we observe convergence in expectation, also the Ky Fan distance will tend to 0. But can this statement be quantified?

For "constant" random variables, i.e., the deterministic case, it turns out that the convergence rate in the Ky Fan metric is the same as the rate observed for the expected value. Furthermore, for the examples in this note we find that the convergence rate in the Ky Fan metric is always slower than the rate observed for the expected value (if the expected value converges).[5]

A tool to quantify these statements is MARKOV's inequality ([21], cf. [10]).

Theorem 3.1 (MARKOV). *For any non-negative random variable X and $c > 0$ we have the estimate*

$$\mathbb{P}(X \geq c) \leq \frac{\mathbb{E}(X)}{c}. \tag{9}$$

From this theorem we obtain, by setting $X = \|x\|^s$ with $s > 0$

$$\mathbb{P}(\|x\| \geq c) = \mathbb{P}(\|x\|^s \geq c^s) \leq \frac{\mathbb{E}(\|x\|^s)}{c^s}. \tag{10}$$

Using this inequality, we can determine bounds on the Ky Fan distance in terms of the expected value as follows.

Theorem 3.2. *Let μ_1 and μ_2 be the distributions of two random variables ξ_1 and ξ_2. The Ky Fan and the Prokhorov distance of the*

---

[5] See also the results in [18, Ch. 5.6].

distributions can be bounded via the expected value of the distance of the random variables as follows

$$\rho_P(\mu_1, \mu_2) \leq \rho_K(\xi_1, \xi_2) \leq \sqrt{\mathbb{E}(\|\xi_1 - \xi_2\|)}.$$

In particular, for arbitrary $s > 0$ (with possibly infinite right-hand side)

$$\rho_P(\mu_1, \mu_2) \leq \rho_K(\xi_1, \xi_2) \leq \mathbb{E}(\|\xi_1 - \xi_2\|^s)^{1/(s+1)}. \tag{11}$$

Proof. Due to (10) we have

$$\mathbb{P}(\|\xi_1 - \xi_2\| \geq c) \leq \frac{\mathbb{E}(\|\xi_1 - \xi_2\|^s)}{c^s}.$$

Solving the equation $c = \mathbb{E}(\|\xi_1 - \xi_2\|^s)/c^s$ concludes the proof. □

Remark 3.3. The bound in (11) leads to the following expectations:

- If ξ_1, ξ_2 are deterministic quantities, the right-hand side of (11) converges to $\|\xi_1 - \xi_2\|$ for $s \to \infty$, and we obtain that $\rho_P(\mu_1, \mu_2)$ and $\rho_K(\xi_1, \xi_2)$ converge at least as fast as the expected value. (As can be seen from (1) and (2) here even $\rho_P(\mu_1, \mu_2) = \rho_K(\xi_1, \xi_2) = \|\xi_1 - \xi_2\|$.)
- If the probability for large values of ξ_1, ξ_2 decays exponentially fast (as it is for instance the case for Gaussian random variables), all moments are finite, but they will grow with s. The relation from $\rho_P(\mu_1, \mu_2)$ and $\rho_K(\xi_1, \xi_2)$ to $\mathbb{E}(\|\xi_1 - \xi_2\|)$ becomes logarithmic (cf. [18, Ch. 6.2]).
- Finally, if not infinitely many moments are finite, but only moments up to some $s_0 < \infty$, the speeds in the two concepts can show significant differences (cf. the previous section and [18, Ch. 5]). This is for instance the case when ξ_1, ξ_2 are decaying polynomially, or come from a Lévy-distribution ([4, 24]).

In this work, we investigated properties of the Prokhorov and the Ky Fan metric, and pointed out connections and differences to other measures of convergence. For the Prokhorov metric, convergence (with rates) of Tikhonov regularized solutions of stochastic linear ill-posed problems was investigated in [13]. In [18] as well as in some forthcoming papers, these results of [13] have been extended to nonlinear problems, to more general regularization methods, and to the Ky Fan metric.

Finally, it should be mentioned that these concepts will also allow to obtain quantitative convergence results for estimators obtained by the Bayesian approach of inverse problems (see [20]).

Acknowledgments

The author thanks Prof. H. W. ENGL and S. KINDERMANN for useful and stimulating discussions.

This work has been supported by the Austrian National Science Foundation FWF under project grant SFB F013/F1308.

References

[1] BANKS, H. T., BIHARI, K. L. (2001) Modelling and estimating uncertainty in parameter estimation. Inverse Problems **17**: 95–111
[2] BARTLE, R. G. (1966) The Elements of Integration. Wiley, New York
[3] BAUER, H. (1972) Probability Theory and Elements of Measure Theory. Holt, Rinehart and Winston, Orlando, Florida
[4] BERTOIN, J. (1996) Lévy Processes (Cambridge Tracts in Mathematics, Vol. 121). Cambridge University Press, Cambridge
[5] BILLINGSLEY, P. (1968) Convergence of Probability Measures. Wiley, New York
[6] BISSANTZ, N., HOHAGE, T., MUNK, A. (2004) Consistency and rates of convergence of nonlinear Tikhonov regularization with random noise. Inverse Problems **20**: 1773–1789
[7] DAVIDSON, J. (1994) Stochastic Limit Theory. The Clarendon Press/Oxford University Press, New York
[8] DUDLEY, R. M. (1968) Distances of probability measures and random variables. Ann. Math. Stat. **39**: 1563–1572
[9] DUDLEY, R. M. (2003) Real Analysis and Probability, 2nd ed. Cambridge University Press, Cambridge, UK
[10] EISENBERG, B., GHOSH, B. (2001) A generalization of Markov's inequality. Stat. Probab. Lett. **53**: 59–65
[11] ELSTRODT, J. (1999) Maß- und Integrationstheorie (Measure and Integration Theory), 2., korr. Aufl. Springer, Berlin Heidelberg New York
[12] ENGL, H. W., HANKE, M., NEUBAUER, A. (1996) Regularization of Inverse Problems. Kluwer, Dordrecht
[13] ENGL, H. W., HOFINGER, A., KINDERMANN, S. (2005) Convergence rates in the Prokhorov metric for assessing uncertainty in ill-posed problems. Inverse Problems **21**: 399–412
[14] ENGL, H. W., WAKOLBINGER, A. (1983) On weak limits of probability distributions on Polish spaces. Stochastic Anal. Appl. **1**: 197–203
[15] ENGL, H. W., WAKOLBINGER, A. (1995) Continuity properties of the extension of a locally Lipschitz continuous map to the space of probability measures. Monatsh. Math. **100**: 85–103
[16] FAN, K. (1944) Entfernung zweier zufälliger Größen und die Konvergenz nach Wahrscheinlichkeit. Mathematische Zeitschrift **49**: 681–683
[17] GIBBS, A. L., SU, F. E. (2002) On choosing and bounding probability metrics. Int. Stat. Rev. **70**: 419–435
[18] HOFINGER, A. (2006) Ill-Posed Problems: Extending the Deterministic Theory to a Stochastic Setup (Schriften der Johannes-Kepler-Universität Linz, Reihe C: Technik und Naturwissenschaften, Vol. 51). Trauner, Linz (Thesis)
[19] HUBER, P. (1981) Robust Statistics. Wiley, New York
[20] KAIPIO, J., SOMERSALO, E. (2005) Statistical and Computational Inverse Problems (Applied Mathematical Sciences, Vol. 160). Springer, New York

[21] MARKOV, A. A. (1913) Ischislenie Veroiatnostei (Calculus of Probabilities), 3rd ed. Gosizdat, Moscow
[22] PARTHASARATHY, K. R. (1967) Probability Measures on Metric Spaces. Academic Press, New York
[23] PROKHOROV, Y. V. (1956) Convergence of random processes and limit theorems in probability theory. Theor. Prob. Appl. **1**: 157–214
[24] SATO, K.-I. (1999) Lévy Processes and Infinitely Divisible Distributions. (Cambridge Studies in Advanced Mathematics, Vol. 68). Cambridge University Press, Cambridge
[25] SCHAY, G. (1974) Nearest random variables with given distributions. Ann. Probab. **2**: 163–166
[26] SMALE, S., ZHOU, D.-X. (2004) Shannon sampling and function reconstruction from point values. Bull. Amer. Math. Soc. **41**: 279–305
[27] STRASSEN, V. (1965) The existence of probability measures with given marginals. Ann. Math. Stat. **36**: 423–439
[28] WERNER, D. (1995) Funktionalanalysis (Functional Analysis). Springer, Berlin Heidelberg New York
[29] WHITT, W. (1974) Preservation of rates of convergence under mappings. Z. Wahrscheinlichkeitstheor. **29**: 39–44
[30] YOSIDA, K. (1980) Functional Analysis, 6th ed. Springer, Berlin Heidelberg New York

Author's address: Dr. Andreas Hofinger, Spezialforschungsbereich SFB F013/F1308, Johann Radon Institute for Computational and Applied Mathematics, Austrian Academy of Sciences, A-4040 Linz, Austria. E-Mail: andreas.hofinger@oeaw.ac.at.

Sitzungsber. Abt. II (2006) 215: 127–137

Sitzungsberichte
Mathematisch-naturwissenschaftliche Klasse Abt. II
Mathematische, Physikalische und Technische Wissenschaften

© Österreichische Akademie der Wissenschaften 2007
Printed in Austria

Stability of the Translation Equation in Rings of Formal Power Series and Partial Extensibility of One-Parameter Groups of Truncated Formal Power Series

By

Wojciech Jabłoński and **Ludwig Reich**

(Vorgelegt in der Sitzung der math.-nat. Klasse am 12. Oktober 2006
durch das w. M. Ludwig Reich)

Abstract

We prove in this paper that stability of the translation equation in the ring of formal power series $\mathbb{K}[[X]]$ (more precisely, in the group of invertible formal series over $\mathbb{K}$) is equivalent to some kind of extensibility of one-parameter groups of truncated formal power series. From this we deduce the stability of the translation equation in $\mathbb{K}[[X]]$.

AMS Subject Classification: 39B82, 39B12, 13F25.
Key words and phrases: Formal power series, translation equation, stability.

1. Introduction

Following the famous problem of the stability of the equation of homomorphism posed by S. ULAM (cf. [8]) many authors have studied such problems for various functional equations (see [1] and [2]). We are going to consider here the problem of stability for the translation equation in the ring of formal power series in one indeterminate. Surprisingly, the considered problem is close to the problem of extensibility of one-parameter groups of truncated formal power series.

In fact, we will show that the stability of the translation equation in the ring of formal power series is equivalent to the possibility of extension of some restriction of a given truncated formal power series (for the concepts of truncation and extension we refer the reader to [5]).

Our main results about the stability of the translation equation (1) in a ring $\mathbb{K}[[X]]$ of formal power series with $\mathbb{K} \in \{\mathbb{R}, \mathbb{C}\}$ are Theorems 2 and 3 stating Hyers-Ulam stability of the translation equation under a rather mild condition on the group G of the "time parameter" t. Let us also mention that the constants in these stability results are best possible.

In this introduction we recall basic notions about formal series and their truncations. Let $\mathbb{N}$ stand for the set of all positive integers and let $\mathbb{K}$ be the field of real or complex numbers. For $k, l \in \mathbb{N}$, by $|k, l|$ we denote the set of all integers n such that $k \leq n \leq l$. Similarly, by $|k, \infty|$ we mean the set of all integers n with $n \geq k$. Throughout this paper we will assume that $|\infty, \infty| = \emptyset$, $\sum_{t \in \emptyset} a_t = 0$ and $\prod_{t \in \emptyset} a_t = 1$.

By $\mathbb{K}[[X]]$ we denote the ring of all power series $\sum_{i=0}^{\infty} c_i X^i$ with coefficients in $\mathbb{K}$. For any formal power series $p(X) = \sum_{i=0}^{\infty} c_i X^i$ we define

$$\operatorname{ord} p(X) := \min\{i \in \mathbb{N} \cup \{0\}: c_i \neq 0\}$$

assuming additionally that $\min \emptyset = \infty$.

The function $d: \mathbb{K}[[X]] \times \mathbb{K}[[X]] \to [0, \infty)$,

$$d(p_1(X), p_2(X)) = (1 + \operatorname{ord}(p_1(X) - p_2(X)))^{-1},$$

properly defines a metric and $(\mathbb{K}[[X]], d)$ is a complete metric space. Moreover, the inequality $d(p_1(X), p_2(X)) < \varepsilon$ is equivalent with $\operatorname{ord}(p_1(X) - p_2(X)) > (1/\varepsilon) - 1$.

A formal power series $p(X) \in \mathbb{K}[[X]]$ can be *substituted into the series* $q(X) = \sum_{i=0}^{\infty} c_i X^i$ provided $\operatorname{ord} p(X) \geq 1$ (only in this case we can properly compute the power series $(q \circ p)(X) := \sum_{i=0}^{\infty} c_i (p(X))^i$). Denote $\Gamma(\mathbb{K}) := \{p(X) \in \mathbb{K}[[X]]: \operatorname{ord} p(X) = 1\}$. Then $\Gamma(\mathbb{K})$ with the substitution as a binary operation is a group.

Finally we will need the notion of congruence modulo X^{s+1} and the notion of truncated formal power series. It is easy to see that for $s \in \mathbb{N}$ the set

$$I_s = \{p(X) \in \mathbb{K}[[X]]: \operatorname{ord} p(X) \geq s + 1\} = X^{s+1} \mathbb{K}[[X]]$$

forms an ideal in the ring $\mathbb{K}[\![X]\!]$. We define *a congruence modulo X^{s+1}* in the following way. We say that for $p_1(X), p_2(X) \in \mathbb{K}[\![X]\!]$ we have

$$p_1(X) \equiv p_2(X) \bmod X^{s+1} \iff p_1(X) - p_2(X) \in I_s.$$

This clearly means that X^{s+1} is a divisor of $(p_1 - p_2)(X) :=$ $p_1(X) - p_2(X)$. We consider the quotient ring $\mathbb{K}[\![X]\!]/I_s$ of all cosets $[p(X)]_s = p(X) + I_s$. With every coset $p(X) + I_s$, where $p(X) = \sum_{i=0}^{\infty} c_i X^i \in \mathbb{K}[\![X]\!]$, we will associate *an s-truncation of a formal power series $p(X)$* defined by

$$p^{[s]}(X) := \sum_{i=0}^{s} c_i X^i \in \mathbb{K}[\![X]\!]_s \subset \mathbb{K}[\![X]\!].$$

In the set $\mathbb{K}[\![X]\!]_s$ we introduce, in a natural way, an addition, a multiplication and a substitution as follows: For $p(X), q(X) \in \mathbb{K}[\![X]\!]_s$ let

$$(p+q)(X) = (p+q)^{[s]}(X),$$
$$(pq)(X) := (pq)^{[s]}(X),$$

and, in the case when ord $q(X) \geq 1$,

$$(p \circ q)(X) := (p \circ q)^{[s]}(X).$$

Then $(\mathbb{K}[\![X]\!]_s, +, \cdot)$ is a ring which is isomorphic with $\mathbb{K}[\![X]\!]/I_s$. Moreover, the set $\Gamma(\mathbb{K})_s := \{p(X) \in \mathbb{K}[\![X]\!]_s \colon \operatorname{ord} p(X) = 1\}$ is a group under substitution.

Definition 1. By a one-parameter group of formal power series we mean a homomorphism $\Phi \colon G \to \Gamma(\mathbb{K})$ from a group $(G, +)$ into $(\Gamma(\mathbb{K}), \circ)$.

For a one-parameter group of formal power series $\Phi \colon G \to \Gamma(\mathbb{K})$ let us denote $F_t(X) = F(t, X) = \Phi(t)(X) = \sum_{i=1}^{\infty} c_i(t) X^i$, where $c_1 \colon G \to \mathbb{K} \setminus \{0\}$, $c_i \colon G \to \mathbb{K}$ for $i \geq 2$. Then the family $(F(t, X))_{t \in G}$ satisfies the well-known translation equation in the ring of formal power series, i.e.

$$F(t_1 + t_2, X) = F(t_1, F(t_2, X)) \qquad \text{for} \quad t_1, t_2 \in G, \qquad (1)$$

or, briefly,

$$F_{t_1 + t_2} = F_{t_1} \circ F_{t_2} \qquad \text{for} \quad t_1, t_2 \in G.$$

Analogously, we can define a one-parameter group of *s*-truncated formal power series as a homomorphism $\Phi_s \colon G \to \Gamma(\mathbb{K})_s$. If we denote $F_t^{[s]}(X) = F^{[s]}(t, X) = \Phi_s(t)(X) = \sum_{i=1}^{s} c_i(t) X^i$, where $c_1 \colon G \to \mathbb{K} \setminus \{0\}$,

$c_i \colon G \to \mathbb{K}$ for $2 \leq i \leq s$, then the family $(F^{[s]}(t, X))_{t \in G}$ satisfies also the translation equation

$$F^{[s]}(t_1 + t_2, X) = F^{[s]}(t_1, F^{[s]}(t_2, X)) \bmod X^{s+1} \qquad \text{for} \quad t_1, t_2 \in G, \quad (2)$$

in the ring of s-truncated formal power series.

Definition 7 contains the notion of stability we are investigating in this paper. It is, in fact, the notion of stability which is used in the theory of the classical Cauchy equation, since by the definition of the metric d Definition 7 can easily be reformulated in terms of this metric. However, for considerations involving formal series the concept of order seems to be more convenient than the related metric d.

2. Stability and Extensibility

We will show here that the Hyers-Ulam stability of the translation equation (1) in formal power series is equivalent with some special kind of extensibility of one-parameter groups of truncated formal power series.

Let us begin with the problem of extensibility of one-parameter groups of s-truncated formal power series (cf. [7]).

Definition 2. Let $(G, +)$ be a group. A one-parameter group $(F_t^{[s]}(X))_{t \in G}$ of s-truncated formal power series is called *extensible* provided there exists a one-parameter group of $(s + 1)$-truncated formal power series $(\overline{F}_t^{[s+1]}(X))_{t \in G}$ such that

$$\left(\overline{F}_t^{[s+1]}\right)^{[s]}(X) = F_t^{[s]}(X) \qquad \text{for every} \quad t \in G.$$

In other words, we ask whether for a one-parameter group of s-truncated formal power series $F^{[s]}(t, X) = \sum_{i=1}^s c_i(t) X^i$ there exists a function $c_{s+1} \colon G \to \mathbb{K}$ such that $\overline{F}^{[s+1]}(t, X) = \sum_{i=1}^{s+1} c_i(t) X^i$ is a one-parameter group of $(s + 1)$-truncated formal power series.

The partial solution of this problem one can find in [4, 5, 6]. Here we will need another kind of extensibility. Namely

Definition 3. Let $r \in (0, 1]$ and let $(G, +)$ be a group. A one-parameter group $F^{[s]}(t, X) = \sum_{i=1}^s c_i(t) X^i$, $t \in G$ of s-truncated formal power series is called *r-partially ∞-extensible* if there exists a one-parameter group of formal power series $\overline{F}(t, X) = \sum_{i=1}^\infty \bar{c}_i(t) X^i$, $t \in G$ such that

$$(\overline{F})^{[rs]}(t, X) = \left(F^{[s]}\right)^{[rs]}(t, X) \qquad \text{for} \quad t \in G$$

or, equivalently, $c_i = \bar{c}_i$ for $1 \leq i \leq [rs]$.

Definition 4. Let $r \in (0, 1]$ and let $(G, +)$ be a group. The translation equation (1) in a ring of formal power series is called *r-stable* if for every positive integer M and for every family $F(t, X) = \sum_{i=1}^{\infty} c_i(t) X^i$, $t \in G$ of formal power series such that

$$\operatorname{ord}(F(t_1 + t_2, X) - F(t_1, F(t_2, X))) > M \qquad \text{for every} \quad t_1, t_2 \in G,$$

there is a one-parameter group of formal power series $\overline{F}(t, X) = \sum_{i=1}^{\infty} \bar{c}_i(t) X^i$ such that

$$\operatorname{ord}(F(t, X) - \overline{F}(t, X)) > rM \qquad \text{for} \quad t \in G$$

or, equivalently, $c_i = \bar{c}_i$ for $1 \leq i \leq [rM]$.

We prove now that r-partial ∞-extensibility of one-parameter groups of truncated formal power series is equivalent to Hyers-Ulam r-stability of the translation equation (1) in rings of formal power series.

Proposition 1. *Fix $r \in (0, 1]$. The translation equation (1) in the ring of formal power series is r-stable if and only if each one-parameter group of truncated formal power series is r-partially ∞-extensible.*

Proof. Assume that the translation equation in $\mathbb{K}[[X]]$ is r-stable. Fix $s \in \mathbb{N}$ arbitrarily and let $F^{[s]}(t, X) = \sum_{i=1}^{s} c_i(t) X^i$ be a one-parameter group of s-truncated formal power series. Put $F(t, X) = \sum_{i=1}^{\infty} c_i(t) X^i$, where $c_i \colon G \to \mathbb{K}$ for $i > s$ are arbitrary. Then

$$\operatorname{ord}(F(t_1 + t_2, X) - F(t_1, F(t_2, X))) > s \qquad \text{for} \quad t_1, t_2 \in G.$$

Since the translation equation is r-stable, we find a one-parameter group of formal power series $\overline{F}(t, X) = \sum_{i=1}^{\infty} \bar{c}_i(t) X^i$ such that

$$\operatorname{ord}(F(t, X) - \overline{F}(t, X)) > rs \qquad \text{for} \quad t \in G.$$

This means that $c_i = \bar{c}_i$ for $1 \leq i \leq [rs]$, so every one-parameter group of truncated formal power series is r-partially ∞-extensible.

Now, assume that each one-parameter group of truncated formal power series is r-partially ∞-extensible. Fix $M \in \mathbb{N}$ and let a family $F(t, X) = \sum_{i=1}^{\infty} c_i(t) X^i$, $t \in G$ be such that

$$\operatorname{ord}(F(t_1 + t_2, X) - F(t_1, F(t_2, X))) > M \qquad \text{for} \quad t_1, t_2 \in G.$$

Then $F^{[M]}(t, X) = \sum_{i=1}^{M} c_i(t) X^i$ is a one-parameter group of M-truncated formal power series. On account of our assumption there exists a one-parameter group $\overline{F}(t, X) = \sum_{i=1}^{\infty} \bar{c}_i(t) X^i$ such that $c_i = \bar{c}_i$ for $1 \leq i \leq [rM]$. Thus

$$\operatorname{ord}(F(t, X) - \overline{F}(t, X)) > rM \qquad \text{for} \quad t \in G.$$

This finishes the proof. $\qquad\qquad\qquad\qquad\qquad\qquad\qquad\qquad \Box$

Thus the problem of stability of the translation equation (1) can be reduced to discussion of extensibility of one-parameter groups of truncated formal power series. We will need a certain form of the general solution of the translation equations (1)–(2) in $\mathbb{K}[\![X]\!]$. We will assume here that $(G, +)$ is an abelian group such that for every generalized exponential function $f\colon G \to \mathbb{K}\backslash\{0\}$ we have either $f = 1$ or f takes infinitely many values. This is equivalent to the following assumption

(A) If $\mathbb{K} = \mathbb{R}$ then G has no subgroup of index 2 and if $\mathbb{K} = \mathbb{C}$ then G has no subgroup with a finite index distinct form G.

Indeed, for a generalized exponential function $f\colon G \to \mathbb{K}\backslash\{0\}, f \neq 1$ taking finitely many values we have im $f \cong G/\ker f$, which jointly with the form of finite subgroups of $(\mathbb{K}\backslash\{0\}, \cdot)$ gives us the assumption (A). Conversely, if we assume that (A) does not hold, then one can construct a generalized exponential function $f\colon G \to \mathbb{K}\backslash\{0\}, f \neq 1$ which takes finitely many values.

We have

Theorem 1 (cf. [4, 5, 6]). *Let s be a positive integer or $s = \infty$. Assume that $(G, +)$ is an abelian group satisfying assumption* (A). *There exist sequences of polynomials $(L_n^{p+2})_{n \geq p+2}$ and $(P_n)_{n \geq 2}$ such that $F^{[s]}(t, X) = \sum_{i=1}^{s} c_i(t) X^i$, $c_1\colon G \to \mathbb{K}\backslash\{0\}$, $c_n\colon G \to \mathbb{K}$ for $n \in |2, s|$ is a solution of the translation equation ((1) if $s = \infty$ and (2) otherwise) if and only if one of three possibilities holds:*

(a) $c_1 = 1$ *and* $c_n = 0$ *for every* $n \in |2, s|$,

(b) $c_1 = 1$, $c_n = 0$ *for* $n \in |2, p + 1|$, c_{p+2} *is a nonzero additive function and*

$$c_n(t) = h_n c_{p+2}(t) + L_n(c_{p+2}(t), 1, h_{p+3}, \dots, h_{n-p-1}),$$
$$n \in |p + 3, s - p - 1|,$$
$$c_n(t) = a_n(t) + L_n(c_{p+2}(t), 1, h_{p+3}, \dots, h_{n-p-1}), \quad n \in |s - p, s|,$$

where $(h_n)_{n \in |p+3, s-p-1|}$, $h_n \in \mathbb{K}$ for $n \in |p + 3, s - p - 1|$, is an arbitrary sequence of constants and $a_n\colon G \to \mathbb{K}$ for $n \in |s - p, s|$ are arbitrary additive functions,

(c) $c_1 \neq 1$ *is a generalized exponential function and*

$$c_n(t) = c_1(t)(\lambda_n(c_1(t)^{n-1} - 1) + P_n(c_1(t); \lambda_2, \dots, \lambda_n)), \quad n \in |2, s|,$$

where $(\lambda_n)_{n \geq 2}$, $\lambda_n \in \mathbb{K}$ for $n \geq 2$, is an arbitrary sequence of constants.

Now we will consider the problem of extensibility of one-parameter groups of s-truncated formal power series. In fact, we will discuss a more complicated version of extensibility.

Definition 5 (cf. [7]). Fix $l \in \mathbb{N}$. A one-parameter group $(F_t^{[s]}(X))_{t \in G}$ of s-truncated formal power series is called *l-extensible* if there exists a one-parameter group of $(s+l)$-truncated formal power series $(\overline{F}_t^{[s+l]}(X))_{t \in G}$ such that

$$\left(\overline{F}_t^{[s+l]}\right)^{[s]}(X) = F_t^{[s]}(X) \qquad \text{for every} \quad t \in G.$$

Thus we ask whether for a one-parameter group of s-truncated formal power series $F^{[s]}(t, X) = \sum_{i=1}^{s} c_i(t) X^i$ there exist functions $c_{s+1}, \dots,$ $c_{s+l} \colon G \to \mathbb{K}$ such that $\overline{F}^{[s+l]}(t, X) = \sum_{i=1}^{s+l} c_i(t) X^i$ is a one-parameter group of $(s+l)$-truncated formal power series.

Definition 6 (cf. [7]). A one-parameter group $(F_t^{[s]}(X))_{t \in G}$ of s-truncated formal power series is called ∞-*extensible* if there exists a one-parameter group of formal power series $(\overline{F}_t(X))_{t \in G}$ such that

$$(\overline{F}_t)^{[s]}(X) = F_t^{[s]}(X) \qquad \text{for every} \quad t \in G.$$

This means that we ask whether for a one-parameter group of s-truncated formal power series $F^{[s]}(t, X) = \sum_{i=1}^{s} c_i(t) X^i$ we can find functions $c_{s+1}, c_{s+2}, \dots \colon G \to \mathbb{K}$ such that $\overline{F}(t, X) = \sum_{i=1}^{\infty} c_i(t) X^i$ is a one-parameter group of formal power series.

Now we have all tools to discuss the extension problem for one-parameter groups of truncated formal power series. We will generally assume that $(G, +)$ is an abelian group satisfying (A). By $\mathcal{A}(G, \mathbb{K})$ we denote the vector space over $\mathbb{K}$ of all additive functions $a \colon G \to \mathbb{K}$.

We begin with the discussion of the subcase when $\dim_\mathbb{K} \mathcal{A}(G, \mathbb{K}) \leq 1$. From Theorem 1 one can easily derive

Corollary 1. *Assume that* $(G, +)$ *is an abelian group satisfying* (A) *and such that* $\dim_\mathbb{K} \mathcal{A}(G, \mathbb{K}) \leq 1$. *If* $F^{[s]}(t, X) = \sum_{i=1}^{s} c_i(t) X^i$ *for* $t \in G$, *where* $c_1 \colon G \to \mathbb{K} \backslash \{0\}$, $c_i \colon G \to \mathbb{K}$ *for* $2 \leq i \leq s$, *is a one-parameter group of s-truncated formal power series (this means that* $F^{[s]}(t, X)$ *is of the form given in Theorem 1), then* $(F^{[s]}(t, X))_{t \in G}$ *is l- and* ∞-*extensible for any* $l \in \mathbb{N}$.

If we do not assume that $\dim_\mathbb{K} \mathcal{A}(G, \mathbb{K}) \leq 1$ then we have

Corollary 2. *Assume that* $(G, +)$ *is an abelian group satisfying assumption* (A). *Let* $F^{[s]}(t, X) = \sum_{i=1}^{s} c_i(t) X^i$ *for* $t \in G$, *where* $c_1 \colon G \to \mathbb{K} \backslash \{0\}$, $c_i \colon G \to \mathbb{K}$ *for* $2 \leq i \leq s$, *be a one-parameter group of s-truncated* *formal power series (this means that $F^{[s]}(t,X)$ is of the form given in Theorem 1).*

(1) *If $c_1 \neq 1$, then $(F^{[s]}(t,X))_{t \in G}$ is l- and ∞-extensible.*

(2) *If $c_1 = 1$, then $(F^{[s]}(t,X))_{t \in G}$ is l-extensible if and only if there are constants $d_i \in \mathbb{K}$ for $s - p \leq i \leq \min(s - p + l - 1, s)$ such that $a_i = d_i c_{p+2}$ for every $s - p \leq i \leq \min(s - p + l - 1, s)$. Moreover, $(F^{[s]}(t,X))_{t \in G}$ is ∞-extensible if and only if there are constants $d_i \in \mathbb{K}$ for $s - p \leq i \leq s$) such that $a_i = d_i c_{p+2}$ for every $s - p \leq i \leq s$.*

Remark 1. Note that Corollary 2 states that in the case $c_1 = 1$ a one-parameter group of truncated formal power series $F^{[s]}(t,X) = \sum_{i=1}^{s} c_i(t)X^i$ is ∞-extensible if and only if the coefficient functions c_n for $n \in |s - p, s|$ depend only (cf. Theorem 1(b)) on c_{p+2} and on some constants, but not on another additive function a_n.

We will finish this section with the following very useful and very simple result, which we are able to derive from Corollary 2.

Corollary 3. *Let $F^{[s]}(t,X) = \sum_{i=1}^{s} c_i(t)X^i$ for $t \in G$, where $c_1: G \to \mathbb{K}\backslash\{0\}$, $c_i: G \to \mathbb{K}$ for $2 \leq i \leq s$, be a one-parameter group of s-truncated formal power series. Then the truncated one-parameter group*

$$\left(\left(F^{[s]} \right)^{[\frac{1}{2}s]+1} (t,X) \right)_{t \in G}$$

is ∞-extensible.

Proof. Assume that $F^{[s]}(t,X) = \sum_{i=1}^{s} c_i(t)X^i$ is a one-parameter group of s-truncated formal power series. The case $c_1 \neq 1$ is trivial. Assume that $c_1 = 1$. Then (see Theorem 1(b)) the coefficient functions c_n with $n \in |p + 2, [\frac{1}{2}s] + 1|$ depend only on c_{p+2}, but not on another additive function a_n. $\qquad \square$

3. Stability of the Translation Equation in $\mathbb{K}[\![X]\!]$

Now we prove results stating Hyers-Ulam stability of the translation equation in a ring of formal power series. We consider a definition of stability of functional equations, which is more general than r-stability.

Definition 7. The translation equation (1) in a ring of formal power series is called stable if for every positive integer N we find a positive integer M such that for every family $F(t,X) = \sum_{i=1}^{\infty} c_i(t)X^i$, $t \in G$ of formal power series such that

$$\operatorname{ord}(F(t_1 + t_2, X) - F(t_1, F(t_2, X))) > M \qquad \text{for every} \quad t_1, t_2 \in G,$$

there is a one-parameter group of formal power series $\overline{F}(t,X) = \sum_{i=1}^{\infty} \bar{c}_i(t)X^i$ such that

$$\text{ord}(F(t,X) - \overline{F}(t,X)) > N \qquad \text{for} \quad t \in G$$

or, equivalently, $c_i = \bar{c}_i$ for $1 \le i \le N$.

We will assume that $(G, +)$ is an abelian group satisfying (A). First let us assume additionally that $\dim_{\mathbb{K}} \mathcal{A}(G, \mathbb{K}) \le 1$. Then, on account of Corollary 1, we have

Theorem 2. *Assume that $(G, +)$ is an abelian group satisfying* (A) *and such that $\dim_{\mathbb{K}} \mathcal{A}(G, \mathbb{K}) \le 1$. Let $M \in \mathbb{N}$ be fixed. If $(F(t,x))_{t \in G}$ is a family of formal power series such that*

$$\text{ord}(F(t_1 + t_2, X) - F(t_1, F(t_2, X))) > M \qquad \text{for every} \quad t_1, t_2 \in G, \tag{3}$$

then there exists a one-parameter group of formal power series $(\overline{F}(t,X))_{t \in G}$ such that

$$\text{ord}(F(t,X) - \overline{F}(t,X)) > M \qquad \text{for} \quad t \in G. \tag{4}$$

Proof. Note that the condition (3) means that for a family $(F(t,X))_{t \in G}$ with $F(t,X) = \sum_{i=1}^{\infty} c_i(t)X^i$, its M-truncation $(F^{[M]}(t,X))_{t \in G}$ satisfies (2) with $s = M$. Then on account of Corollary 1 we know that $(F(t,X))_{t \in G}$ is ∞-extensible. This means that we find a one-parameter group of formal power series $(\overline{F}(t,X))_{t \in G}$ satisfying (4). □

If we do not assume that $\dim_{\mathbb{K}} \mathcal{A}(G, \mathbb{K}) \le 1$, we have

Theorem 3. *Assume that $(G, +)$ is an abelian group satisfying* (A). *Let $M \in \mathbb{N}$ be fixed. If $(F(t,x))_{t \in G}$ is a family of formal power series satisfying* (3), *then there exists a one-parameter group of formal power series $(\overline{F}(t,X))_{t \in G}$ such that*

$$\text{ord}(F(t,X) - \overline{F}(t,X)) > \left[\frac{1}{2}M\right] + 1 \qquad \text{for} \quad t \in G. \tag{5}$$

Proof. As in the proof of Theorem 2, the condition (3) means that for a family $(F(t,X))_{t \in G}$ with $F(t,X) = \sum_{i=1}^{\infty} c_i(t)X^i$, its M-truncation $(F^{[M]}(t,X))_{t \in G}$ satisfies (2) with $s = M$. Then on account of Corollary 3 we know that the truncation

$$\left(\left(F^{[M]}\right)^{[\frac{1}{2}M]+1}(t,X) \right)_{t \in G}$$

is ∞-extensible. This means that we find a one-parameter group of formal power series $(\overline{F}(t,X))_{t \in G}$ such that (5) holds. □

Remark 2. Let us note that the one-parameter group of formal power series $(\overline{F}(t, X))_{t \in G}$ satisfying (4) and (5), respectively, which exists by Theorems 2 and 3 is not unique (cf. Theorem 1). This follows from the fact that constants h_n and λ_n, respectively, can be chosen arbitrarily for $n > M$.

Remark 3. The estimations (4) and (5) obtained in the Theorems 2 and 3 are the best possible. First let us consider the case when $\dim_{\mathbb{K}} \mathcal{A}(G, \mathbb{K}) \leq 1$. Fix $M \in \mathbb{N}$ arbitrarily. Let $c_1 = 1$, $c_i = 0$ for $i \in |2, M|$, $c_{M+1} \colon G \to \mathbb{K}$ let be an arbitrary function which is not additive and take arbitrary functions $c_i \colon G \to \mathbb{K}$ for $i \in |M + 2, \infty|$. Put $F(t, X) = \sum_{i=1}^{\infty} c_i(t) X^i$. It is clear (cf. Theorem 1) that

$$\mathrm{ord}(F(t_1 + t_2, X) - F(t_1, F(t_2, X))) = M + 1 \qquad \text{for} \quad t_1, t_2 \in G.$$

Let $\bar{c}_1 = 1$ and $\bar{c}_i = 0$ for every $i \in |2, \infty|$. Then $\overline{F}(t, X) = \sum_{i=1}^{\infty} \bar{c}_i(t) X^i$ is a one-parameter group of formal power series such that

$$\mathrm{ord}(F(t, X) - \overline{F}(t, X)) = M + 1 \qquad \text{for} \quad t \in G.$$

Now assume that $\dim_{\mathbb{K}} \mathcal{A}(G, \mathbb{K}) \geq 2$. Fix $M \in \mathbb{N}$ arbitrarily. In the cases $M \in \{1, 2\}$ works the same example as above. So assume that $M \geq 3$ and put $p := [M/2] - 1$. Let $c_1 = 1$, $c_i = 0$ for $i \in |2, p + 1|$ and let $c_{p+2} \colon G \to \mathbb{K}$ be a nonzero additive function. Furthermore, let

$$c_i(t) = h_i c_{p+2}(t) + L_i(c_{p+2}(t), h_{p+3}, \ldots, h_{i-p-1}),$$
$$i \in |p + 3, M - p - 1|,$$
$$c_i(t) = a_i(t) + L_i(c_{p+2}(t), h_{p+3}, \ldots, h_{i-p-1}), \qquad i \in |M - p, M|,$$

where $(L_n^{p+2})_{n \geq p+2}$ are the polynomials from Theorem 1, $(h_n)_{n \in |p+3, M-p-1|}$, $h_n \in \mathbb{K}$ for $n \in |p + 3, M - p - 1|$, is a fixed sequence of constants and $a_n \colon G \to \mathbb{K}$ for $n \in |s - p, s|$ are fixed additive functions such that $a_{M-p} \neq b c_{p+2}$ for any $b \in \mathbb{K}$ (this is possible because of the assumption $\dim_{\mathbb{K}} \mathcal{A}(G, \mathbb{K}) \geq 2$). Finally, let c_i for $i \geq M + 1$ be arbitrary. Put $F(t, X) = \sum_{i=1}^{\infty} c_i(t) X^i$. It is clear (cf. Theorem 1) that

$$\mathrm{ord}(F(t_1 + t_2, X) - F(t_1, F(t_2, X))) = M + 1 \qquad \text{for} \quad t_1, t_2 \in G,$$

because $a_{M-p} \neq b c_{p+2}$ for any $b \in \mathbb{K}$. Let $\bar{c}_i = c_i$ for every $i \in |1, M - p - 1|$ and put

$$\bar{c}_i(t) = h_i c_{p+2}(t) + L_i(c_{p+2}(t), h_{p+3}, \ldots, h_{i-p-1})$$
$$\text{for} \quad i \in |M - p - 1, \infty|.$$

Then $\overline{F}(t,X) = \sum_{i=1}^{\infty} \overline{c}_i(t)X^i$ is a one-parameter group of formal power series such that

$$\mathrm{ord}(F(t,X) - \overline{F}(t,X)) = M - p = \left[\frac{M}{2}\right] + 1 \qquad \text{for} \quad t \in G.$$

References

[1] FORTI, G. L. (1995) Hyers-Ulam stability of functional equations in several variables. Aequationes Math. **50**: 143–190

[2] HYERS, D. H., ISAC, G., RASIAS, TH. M. (1998) Stability of Functional Equations in Several Variables (Progress in Nonlinear Differential Equations and Their Applications, Vol. 34). Birkhäuser, Boston Basel Berlin

[3] JABŁOŃSKI, W. (1999) On extensibility of some homomorphisms. Rocznik Nauk.-Dydak. WSP w Krakowie **16/207**: 35–43

[4] JABŁOŃSKI, W., REICH, L. (2005) On the solutions of the translation equation in rings of formal power series. Abh. Math. Sem. Univ. Hamburg **75**: 179–201

[5] JABŁOŃSKI, W., REICH, L. (2005) On the form of homomorphisms into the differential group L_s^1 and their extensibility. Result. Math. **47**: 61–68

[6] JABŁOŃSKI, W., REICH, L.: On homomorphisms of an Abelian group into the group of invertible formal power series (manuscript)

[7] REICH, L. (1991) Problem. Aequationes Math. **41**: 248–310

[8] ULAM, S. M. (1960) A Collection of Mathematical Problems. Interscience, New York London

Authors' addresses: Dr. Wojciech Jabłoński, Department of Mathematics, University of Rzeszów, Rejtana 16 A, 35-310 Rzeszów, Poland. E-Mail: wojciech@univ.rzeszow.pl; Prof. Dr. Ludwig Reich, Institute of Mathematics, Karl-Franzens-University Graz, Heinrichstrasse 36, 8010 Graz, Austria. E-Mail: ludwig.reich@uni-graz.at.

Sitzungsber. Abt. II (2006) 215: 139–153

Sitzungsberichte
Mathematisch-naturwissenschaftliche Klasse Abt. II
Mathematische, Physikalische und Technische Wissenschaften

Remarks on Some Sequences of Binomial Sums

By

Johann Cigler

(Vorgelegt in der Sitzung der math.-nat. Klasse am 16. November 2006
durch das w. M. Johann Cigler)

Abstract

We give simple proofs for the recurrence relations of some sequences of binomial sums which have previously been obtained by other more complicated methods.

Mathematics Subject Classification (2000): 05A10, 11B39, 39A10.
Key words: Recurrence, Lucas polynomial, Fibonacci polynomial, binomial coefficient.

1. Introduction

Modifying an idea of BRIETZKE [2] we give simple proofs for the recurrence relations of sequences of binomial sums of the form

$$a(n, m, k, z) = \sum_{j \in \mathbb{Z}} z^j \left(\left\lfloor \dfrac{n}{\dfrac{n - mj + k}{2}} \right\rfloor \right),$$

which have been obtained by other methods in [3].

In order to motivate the method we consider first the well-known special case

$$a(n, 5, k, -1) = \sum_{j \in \mathbb{Z}} (-1)^j \left(\left\lfloor \dfrac{n}{\dfrac{n - 5j + k}{2}} \right\rfloor \right) = (-1)^k \sum_j t(n, k - 5j),$$

with

$$t(n,k) = (-1)^k \left(\left\lfloor \begin{array}{c} n \\ \dfrac{n+k}{2} \end{array} \right\rfloor \right).$$

We use the fact that $t(n,k) = -t(n-1,k-1) - t(n-1,k+1)$ with $t(0,0) = 1, t(0,1) = -1$ and $t(0,k) = 0$ for all other $k \in \mathbb{Z}$.

Define the operator K by $Kf(n,k) = f(n,k-1)$ and the operator N by $Nf(n,k) = f(n+1,k)$. Then

$$t(n) = Nt(n-1) = -(K + K^{-1})t(n-1) = (-1)^n (K + K^{-1})^n t(0).$$

Let $s(n,k)$ on $\mathbb{N} \times \mathbb{Z}$ be the function which satisfies the same recurrence with initial values $s(0,k) = [k = 0]$. Then we have $t(0) = (1-K)s(0)$. Since K is a linear operator we also have $t(n) = (1-K)s(n)$.

Let $\mathcal{F}$ be the vector space of all functions on $\mathbb{N} \times \mathbb{Z}$ which are finite linear combinations of functions $K^j s$, $j \in \mathbb{Z}$. For $f \in \mathcal{F}$ we have $Nf = -(K + K^{-1})f$.

Let T be the linear operator on $\mathcal{F}$ defined by

$$Tf = N^2 f - Nf - f = (K + K^{-1})^2 f + (K + K^{-1})f - f$$
$$= (K^{-2} + K^{-1} + 1 + K + K^2)f.$$

Then

$$\sum_{j \in \mathbb{Z}} K^{5j} T K^i s(0) = \sum_{j \in \mathbb{Z}} K^j s(0) = 1 \qquad \text{for all} \quad i \in \mathbb{Z}$$

since $KT = TK$.

Furthermore

$$\sum_{j \in \mathbb{Z}} K^{5j} Tt(n) = \sum_{j \in \mathbb{Z}} K^{5j} T(-1)^n (K + K^{-1})^n (1-K)s(0)$$
$$= (-1)^n (K + K^{-1})^n (1-K) \sum_{j \in \mathbb{Z}} K^{5j} Ts(0) = 0.$$

Since

$$a(n,5,k,-1) = (-1)^k \sum_j t(n,k-5j)$$

is a finite sum for each k, the sequence $(a(n,5,k,-1))$ satisfies the recurrence

$$a(n+2,5,k,-1) - a(n+1,5,k,-1) - a(n,5,k,-1) = 0 \quad \text{for} \quad n \geq 0.$$

Since the Fibonacci numbers F_n satisfy the same recurrence with initial values $F_0 = 0$ and $F_1 = 1$, we get the following results (cf. ANDREWS [1]):

Proposition 1. *For $k \equiv 0, 1 \pmod{10}$ the initial conditions are $a(0, 5, k) = a(1, 5, k) = 1$ and therefore $a(n, 5, k) = F_{n+1}$.*

For $k \equiv 2, 9 \pmod{10}$ we have $a(0, 5, k) = 0$, $a(1, 5, k) = 1$ and therefore $a(n, 5, k) = F_n$.

For $k \equiv 3, 8 \pmod{10}$ we get $a(0, 5, k) = a(1, 5, k) = 0$ and therefore $a(n, 5, k) = 0$. Furthermore $a(n, 5, k + 5) = -a(n, 5, k)$.

It is interesting to observe that this result has first been proved by SCHUR [6] in a strengthened version: Let

$$\begin{bmatrix} n \\ k \end{bmatrix} = \frac{(1 - q^{n-k+1}) \cdots (1 - q^n)}{(1 - q) \cdots (1 - q^k)}$$

be a q-binomial coefficient. Then the following polynomial version of the celebrated Rogers-Ramanujan identity

$$\sum_{k=0}^{n} q^{k^2} \begin{bmatrix} n - k \\ k \end{bmatrix} = \sum_{k \in \mathbb{Z}} (-1)^k q^{\frac{k(5k-1)}{2}} \begin{bmatrix} n \\ \left\lfloor \dfrac{n + 5k}{2} \right\rfloor \end{bmatrix}$$

holds, which for $q = 1$ reduces to

$$\sum_{k=0}^{n} \binom{n - k}{k} = F_{n+1} = \sum_{j \in \mathbb{Z}} (-1)^j \left(\left\lfloor \dfrac{n + 5j}{2} \right\rfloor \right).$$

An elementary proof of this q-identity may be found in [5].

2. A Useful Method

After this example let us consider a more general case.

For $a, b \in \mathbb{R}$ let $s_{a,b}$ be the function on $\mathbb{N} \times \mathbb{Z}$ defined by $s_{a,b}(0, k) = [k = 0]$ and the recurrence relation

$$s_{a,b}(n, k) = a s_{a,b}(n - 1, k - 1) + b s_{a,b}(n - 1, k) + a s_{a,b}(n - 1, k + 1). \tag{1}$$

This can be written in the form

$$s_{a,b}(n) = (aK^{-1} + b + aK) s_{a,b}(n - 1) = (aK^{-1} + b + aK)^n s_{a,b}(0).$$

Let $\mathcal{F}$ be the vector space of all functions on $\mathbb{N}$ which are finite linear combinations of functions $K^j s_{a,b}, \ j \in \mathbb{Z}$.

For any polynomial

$$p(x) = \sum_{i=0}^{m} a_i x^i$$

we denote by $p(N)$ the linear operator on $\mathcal{F}$ defined by

$$p(N)f(n) = \sum_{i=0}^{m} a_i f(n+i).$$

Then we have $p(N) = p(aK^{-1} + b + aK)$.

We are looking for an operator $p(N)$ with analogous properties as T had in the above example.

To this end we define a sequence of polynomials

$$p_n(x, a, b) = \sum_{k=0}^{n} p_{n,k}(a, b) x^k$$

by the recurrence

$$p_n(x, a, b) = (x - b)p_{n-1}(x, a, b) - a^2 p_{n-2}(x, a, b) \qquad (2)$$

with initial values $p_0(x, a, b) = 1$ and $p_1(x, a, b) = x + a - b$.

Lemma 1. *For all $k \in \mathbb{Z}$ the following identity holds*

$$p_m(N, a, b)s_{a,b}(0, k) = \sum_{i=0}^{m} p_{m,i}(a, b)s_{a,b}(i, k) = a^m[|k| \leq m]. \qquad (3)$$

Proof. It suffices to show that on $\mathcal{F}$

$$p_m(N, a, b) = a^m \sum_{j=-m}^{m} K^j. \qquad (4)$$

It is immediately verified that (4) is true for $m = 0$ and $m = 1$, since

$$(N + a - b) = (aK + a + aK^{-1}).$$

If (4) has already been shown for $m - 1$ and $m - 2$ we get

$$p_m(N, a, b) = (N - b)p_{m-1}(N, a, b) - a^2 p_{m-2}(N, a, b)$$

$$= a(K + K^{-1})a^{m-1} \sum_{j=-m+1}^{m-1} K^j - a^2 a^{m-2} \sum_{j=-m+2}^{m-2} K^j$$

$$= a^m \sum_{j=-m}^{m} K^j.$$

From (3) we get

$$\sum_{i=0}^{m} p_{m,i}(a,b) \sum_{j\in\mathbb{Z}} s_{a,b}(i, k-(2m+1)j) = a^m \tag{5}$$

for each $k \in \mathbb{Z}$.

Application. As an application we consider for each $m \in \mathbb{N}$ the sequence

$$\begin{aligned} a(n, 2m+1, k, -1) &= \sum_{j\in\mathbb{Z}} (-1)^j \binom{n}{\left\lfloor \frac{n-(2m+1)j+k}{2} \right\rfloor} \\ &= (-1)^k \sum_j t(n, k-(2m+1)j). \end{aligned}$$

As shown above we have $t = (1-K)s_{-1,0}$. Therefore by (5) we get

$$\sum_{i=0}^{m} p_{m,i}(-1,0) a(0, 2m+1, k, -1) = 0.$$

Formula (1) implies that $t(n)$ is a finite linear combination of functions $K^j t(0)$. Therefore we also get

$$\begin{aligned} &p_m(N, -1, 0) a(n, 2m+1, k, -1) \\ &\quad = \sum_{i=0}^{m} p_{m,i}(-1,0) a(n, 2m+1, k, -1) = 0. \end{aligned}$$

Now we look for an explicit expression for $p_n(x, -1, 0)$.

We know that it satisfies the recurrence

$$p_n(x, -1, 0) = x p_{n-1}(x, -1, 0) - p_{n-2}(x, -1, 0)$$

with initial values $p_0(x, -1, 0) = 1$ and $p_1(x, -1, 0) = x - 1$.

Recall that the Fibonacci polynomials

$$\begin{aligned} F_n(x,s) &= \sum_{k=0}^{n-1} \binom{n-1-k}{k} s^k x^{n-2k-1} \\ &= \frac{1}{\sqrt{x^2+4s}} \left(\left(\frac{x+\sqrt{x^2+4s}}{2} \right)^n - \left(\frac{x-\sqrt{x^2+4s}}{2} \right)^n \right) \end{aligned} \tag{6}$$

are characterized by the recurrence

$$F_n(x,s) = xF_{n-1}(x,s) + sF_{n-2}(x,s) \tag{7}$$

with initial conditions $F_0(x,s) = 0$ and $F_1(x,s) = 1$. Therefore

$$p_n(x,-1,0) = F_{n+1}(x,-1) - F_n(x,-1).$$

The first values of the polynomials $p_n(x,-1,0)$ are

$$1, x-1, x^2-x-1, x^3-x^2-2x+1, x^4-x^3-3x^2+2x+1, \ldots.$$

This gives

Theorem 1. *The sequence*

$$a(n,2m+1,k,-1) = \sum_{j\in\mathbb{Z}}(-1)^j\left(\left\lfloor\frac{n-(2m+1)j+k}{2}\right\rfloor\right)$$

satisfies the recurrence relation of order m

$$(F_{m+1}(N,-1) - F_m(N,-1))a(n,2m+1,k,-1) = 0 \tag{8}$$

for each $k\in\mathbb{Z}$.

Remark. This theorem has been proved in [3] with a more complicated method. The recurrence (8) is not for all k the minimal recurrence, because e.g. $a(n,2m+1,m+1,-1) \equiv 0$. But it is so for $a(n,2m+1,0,-1)$, which has a simple combinatorial interpretation. It is the number of the set of all lattice paths in $\mathbb{R}^2$ which start at the origin, consist of $\lfloor\frac{n}{2}\rfloor$ northeast steps $(1,1)$ and $\lfloor\frac{n+1}{2}\rfloor$ southeast steps $(1,-1)$ and which are contained in the strip $-m-1<y<m$ (cf. e.g. [4], [5]).

It is easy to see that the initial values of $a(n,2m+1,0,-1)$ are

$$a(j,2m+1,0,-1) = \left(\left\lfloor\frac{j}{2}\right\rfloor\right) \qquad \text{for} \quad 0\le j<2m.$$

As a special case of Theorem 1 we mention that $a(n,3,0,-1) = 1$. This means

$$\sum_{j\in\mathbb{Z}}(-1)^j\left(\left\lfloor\frac{n-3j}{2}\right\rfloor\right) = 1 \qquad \text{for all} \quad n\in\mathbb{N}.$$

The generating function of the sequence $(a(n, 2m + 1, 0, -1))_{n \geq 0}$ has the form

$$\sum_{n \geq 0} a(n, 2m + 1, 0, -1)x^n = \frac{c_m(x)}{d_m(x)},$$

where

$$d_m(x) = p_m\left(\frac{1}{x}, -1, 0\right)x^m = x^m\left(F_{m+1}\left(\frac{1}{x}, -1\right) - F_{m+1}\left(\frac{1}{x}, -1\right)\right)$$

$$= F_{m+1}(1, -x^2) - xF_m(1, -x^2)$$

and $c_m(x)$ is a polynomial of degree less than m.

The first values of $(c_m(x))_{m \geq 1}$ are

$$c_1(x) = 1, \qquad c_2(x) = 1, \qquad c_3(x) = 1 - x^2,$$
$$c_4(x) = 1 - 2x^2, \qquad c_5(x) = 1 - 3x^2 + x^4, \dots .$$

This suggests that for $m \geq 2$

$$c_m(x) = \sum_{j=0}^{m-1} (-1)^j \binom{m-1-j}{j} x^{2j} = F_m(1, -x^2).$$

This can be proved in the following way: Both $d_m(x)$ and $F_m(1, -x^2)$ satisfy the same recurrence $h_m(x) = h_{m-1}(x) - x^2 h_{m-2}(x)$. This implies that for

$$a_{2m+1}(x) = \sum_{n \geq 0} a(n, 2m + 1, 0, -1)x^n$$

we have

$$d_m(x)a_{2m+1}(x) - d_{m-1}(x)a_{2m-1}(x) + x^2 d_{m-2}(x)a_{2m-3}(x)$$
$$= (d_m(x) - d_{m-1}(x) - x^2 d_{m-2}(x))a_{2m+1}(x) + d_{m-1}(x)(a_{2m+1}(x)$$
$$- a_{2m-1}(x)) + x^2 d_{m-2}(x)(a_{2m+1}(x) - a_{2m-3}(x)).$$

Since the coefficients of x^j for $0 \leq j \leq 2m - 5$ of $a_{2m-3}(x)$ are the same as those of $a_{2m-1}(x)$ and $a_{2m+1}(x)$ we see that for $2m - 4 \geq m - 1$ the polynomial

$$d_m(x)a_{2m+1}(x) - d_{m-1}(x)a_{2m-1}(x) + x^2 d_{m-2}(x)a_{2m-3}(x)$$

which has degree $< m$ must identically vanish. This implies that

$$c_m(x) = d_m(x)a_{2m+1}(x) = F_m(1, -x^2).$$

Corollary 1. *For $m \geq 2$ the generating function for $a(n, 2m+1, 0, -1)$ is given by*

$$\sum_{n \geq 0} a(n, 2m+1, 0, -1)x^n = \frac{F_m(1, -x^2)}{F_{m+1}(1, -x^2) - xF_m(1, -x^2)}. \quad (9)$$

3. A Modification of the Above Method

In order to obtain an analogous result for the sequences $a(n, 2m, k, -1)$ we define a sequence of polynomials

$$q_n(x, a, b) = \sum_{k=0}^{n} q_{n,k}(a, b)x^k$$

by the same recurrence

$$q_n(x, a, b) = (x - b)q_{n-1}(x, a, b) - a^2 q_{n-2}(x, a, b), \quad (10)$$

but with initial values $q_0(x, a, b) = 2$ and $q_1(x, a, b) = x - b$.

Lemma 2. *For all $k \in \mathbb{Z}$ the following identity holds*

$$q_m(N, a, b)s_{a,b}(0, k) = \sum_{i=0}^{m} q_{m,i}(a, b)s_{a,b}(i, k) = a^m[|k| = m]. \quad (11)$$

Proof. It suffices to show that on $\mathcal{F}$

$$q_m(N, a, b) = a^m(K^m + K^{-m}). \quad (12)$$

(12) is true for $m = 0$ and $m = 1$ by inspection.
If it is already shown for $m - 1$ and $m - 2$ we get

$$q_m(N, a, b) = a(K + K^{-1})a^{m-1}(K^{m-1} + K^{-(m-1)})$$
$$- a^2 a^{m-2}(K^{m-2} + K^{-(m-2)}) = a^m(K^m + K^{-m}).$$

Application. As an application let

$$u(n, k) = \left(\begin{array}{c} n \\ \left\lfloor \dfrac{n+k}{2} \right\rfloor \end{array} \right).$$

Then $u(n, k) = u(n-1, k-1) + u(n-1, k+1)$ and $u(0, k) = [k \in \{0, 1\}]$. Therefore

$$u(n, k) = s_{1,0}(n, k) + s_{1,0}(n, k-1) \quad \text{or} \quad u = (1 + K)s_{1,0}.$$

We have

$$a(n,2m,k,-1) = \sum_{j \in \mathbb{Z}} (-1)^j \left(\left\lfloor \frac{n}{\frac{n-(2m)j+k}{2}} \right\rfloor \right)$$

$$= \sum_{j \in \mathbb{Z}} \left(\left(\left\lfloor \frac{n}{\frac{n-(2m)2j+k}{2}} \right\rfloor \right) \right.$$

$$\left. - \left(\left\lfloor \frac{n}{\frac{n-(2m)(2j+1)+k}{2}} \right\rfloor \right) \right)$$

$$= \sum_{j \in \mathbb{Z}} (s_{1,0}(n,k-4mj) - s_{1,0}(n,k-2m-4mj))$$

$$+ \sum_{j \in \mathbb{Z}} (s_{1,0}(n,k-1-4mj) - s_{1,0}(n,k-1-2m-4mj)).$$

Here we get

$$q_m(N,1,0) \sum_{j \in \mathbb{Z}} (s_{1,0}(0,i-4mj) - s_{1,0}(0,i-2m-4mj)) = 0$$

for each i,

because for $i - 4mj = m$ we get $i - 4mj - 2m = -m$ and the sums cancel and for $i - 4mj = -m$ we get $i - 4m(j-1) - 2m = m$. For other values the sum vanishes.

In the same way as above we conclude that

$$q_m(N,1,0) \sum_{j \in \mathbb{Z}} (s_{1,0}(n,i-4mj) - s_{1,0}(n,i-2m-4mj)) = 0$$

too.

In order to give a concrete representation of $q_m(x,1,0)$ recall that the Lucas polynomials

$$L_n(x,s) = \sum_{k=0}^{n-1} \binom{n-k}{k} \frac{n}{n-k} s^k x^{n-2k}$$

$$= \left(\frac{x + \sqrt{x^2+4s}}{2} \right)^n + \left(\frac{x - \sqrt{x^2+4s}}{2} \right)^n \quad (13)$$

are characterized by the recurrence

$$L_n(x,s) = x L_{n-1}(x,s) + s L_{n-2}(x,s) \quad (14)$$

with initial conditions $L_0(x,s) = 2$ and $L_1(x,s) = x$. Therefore $q_n(x,1,0) = L_n(x,-1)$.

The first values of the sequence $(L_n(x,-1))_{n \geq 1}$ are

$$x, \quad x^2 - 2, \quad x^3 - 3x, \quad x^4 - 4x^2 + 2, \ldots .$$

Theorem 2. *For $m \geq 1$ the sequence*

$$a(n, 2m, k, -1) = \sum_{j \in \mathbb{Z}} (-1)^j \left(\left\lfloor \frac{n}{\frac{n - (2m)j + k}{2}} \right\rfloor \right)$$

satisfies the recurrence relation

$$L_m(N, -1)a(n, 2m, k, -1) = 0. \tag{15}$$

Remark. It should be noted that $a(n, 2m, 0, -1)$ has the following combinatorial interpretation. It is the number of the set of all lattice paths in $\mathbb{R}^2$ which start at the origin, consist of $\lfloor \frac{n}{2} \rfloor$ northeast steps $(1,1)$ and $\lfloor \frac{n+1}{2} \rfloor$ southeast steps $(1,-1)$ and which are contained in the strip $-m < y < m$ (cf. e.g. [5]).

The generating function of the sequence $(a(n, 2m, 0, -1))_{n \geq 0}$ is given by

$$\sum_{n \geq 0} a(n, 2m, 0, -1)x^n = \frac{c_m(x)}{d_m(x)},$$

where

$$d_m(x) = q_m\left(\frac{1}{x}, 1, 0\right)x^m = x^m L_m\left(\frac{1}{x}, -1\right) = L_m(1, -x^2)$$

and $c_m(x)$ is a polynomial of degree less than m.

The first values of $(c_m(x))_{m \geq 1}$ are

$$c_1(x) = 1, \quad c_2(x) = 1 + x, \quad c_3(x) = 1 + x - x^2,$$
$$c_4(x) = 1 + x - 2x^2 - x^3, \quad c_5(x) = 1 + x - 3x^2 - 2x^3 + x^4, \ldots .$$

This implies as above that

$$c_m(x) = F_m(1, -x^2) + xF_{m-1}(1, -x^2).$$

Corollary 2. *For $m \geq 2$ the generating function for $a(n, 2m, 0, -1)$ is given by*

$$\sum_{n \geq 0} a(n, 2m, 0, -1)x^n = \frac{F_m(1, -x^2) + xF_{m-1}(1, -x^2)}{L_m(1, -x^2)}. \tag{16}$$

4. Further Applications

4a) The same method can be applied to the general sum

$$a(n,m,k,z) = \sum_{j \in \mathbb{Z}} z^j \left(\begin{array}{c} n \\ \left\lfloor \dfrac{n - mj + k}{2} \right\rfloor \end{array} \right) = \sum_{j \in \mathbb{Z}} z^{2j} \left(\begin{array}{c} n \\ \left\lfloor \dfrac{n - 2mj + k}{2} \right\rfloor \end{array} \right)$$

$$+ \sum_{j \in \mathbb{Z}} z^{2j-1} \left(\begin{array}{c} n \\ \left\lfloor \dfrac{n - 2mj + k + m}{2} \right\rfloor \end{array} \right).$$

Here we get

$$L_m(N, -1)a(0,m,k,z) = L_m(N, -1) \sum_{j \in \mathbb{Z}} z^{2j} u(0, k - 2mj)$$

$$+ L_m(N, -1) \sum_{j \in \mathbb{Z}} z^{2j-1} u(0, k + m - 2mj).$$

In this case we have

$$L_m(N, -1)u(0, k - 2mj) = \begin{cases} 1, & \text{if } k = 2mj - m, \\ 1, & \text{if } k = 2mj + m, \\ 0, & \text{else,} \end{cases}$$

or

$$L_m(N, -1)u(0, k - 2mj) = u(0, k - m - 2mj) + u(0, k + m - 2mj).$$

This implies

$$L_m(N, -1)a(0,m,k,z) = \sum_{j \in \mathbb{Z}} z^{2j} \left(u(0, k - m - 2mj) + u(0, k + m - 2mj) \right)$$

$$+ \sum_{j \in \mathbb{Z}} z^{2j-1} \left(u(0, k + 2m - 2mj) + u(0, k - 2mj) \right)$$

$$= \left(z + \frac{1}{z} \right) a(0,m,k,z).$$

Thus we get

$$\left(L_m(N, -1) - \left(z + \frac{1}{z} \right) \right) a(0, m, k, z) = 0.$$

Theorem 3. *The sequence*

$$a(n,m,k,z) = \sum_{j \in \mathbb{Z}} z^j \left(\begin{array}{c} n \\ \left\lfloor \dfrac{n - mj + k}{2} \right\rfloor \end{array} \right)$$

satisfies the recurrence relation

$$\left(L_m(N,-1) - \left(z + \frac{1}{z} \right) \right) a(n,m,k,z) = 0. \tag{17}$$

Remark. It is easy to see that the initial values of $a(n,m,0,z)$ are

$$a(n,m,0,z) = \begin{pmatrix} j \\ \left\lfloor \dfrac{j}{2} \right\rfloor \end{pmatrix} \qquad \text{for} \quad 0 \le j < m-1,$$

$$a(m-1,m,0,z) = \begin{pmatrix} m-1 \\ \left\lfloor \dfrac{m-1}{2} \right\rfloor \end{pmatrix} + \frac{1}{z},$$

$$a(m,m,0,z) = \begin{pmatrix} m \\ \left\lfloor \dfrac{m}{2} \right\rfloor \end{pmatrix} + \frac{1}{z} + z.$$

The generating function of the sequence $(a(n,m,0,z))$ for $m \ge 1$ has the form

$$\sum_{n \ge 0} a(n,m,0,z)x^n = \frac{c_m(x,z)}{d_m(x,z)}$$

with

$$d_m(x,z) = x^m \left(L_m\left(\frac{1}{x}, -1 \right) - \left(z + \frac{1}{z} \right) \right) = d_m(x) - x^m \left(z + \frac{1}{z} \right)$$

and

$$c_m(x,z) = \frac{x^{m-1}}{z} + F_m(1,-x^2) + x F_{m-1}(1,-x^2).$$

Since $d_m(x) = L_m(1,-x^2)$ and $F_m(1,-x^2) + x F_{m-1}(1,-x^2)$ satisfy the same recurrence $h_m(x) = h_{m-1}(x) - x^2 h_{m-2}(x)$ we get

$$\left(d_m(x) - x^m \left(z + \frac{1}{z} \right) \right) a_m(x) - \left(d_{m-1}(x) - x^{m-1} \left(z + \frac{1}{z} \right) \right) a_{m-1}(x)$$

$$+ x^2 \left(d_{m-2}(x) - x^{m-2} \left(z + \frac{1}{z} \right) \right) a_{m-2}(x)$$

$$= d_{m-1}(x)(a_m(x) - a_{m-1}(x)) + x^2 d_{m-2}(x)(a_m(x) - a_{m-2}(x))$$

$$- x^m \left(z + \frac{1}{z} \right) a_m(x) + x^{m-1} \left(z + \frac{1}{z} \right) a_{m-1}(x) - x^m \left(z + \frac{1}{z} \right) a_{m-2}(x).$$

Since $d_m(0) = 1$ it is easy to verify that for $m \geq 3$

$$d_{m-1}(x)(a_m(x) - a_{m-1}(x)) = -\frac{x^{m-2}}{z} - x^{m-1}z + x^m(\cdots)$$

and

$$x^2 d_{m-2}(x)(a_m(x) - a_{m-2}(x)) = -\frac{x^{m-1}}{z} + x^m(\cdots).$$

Therefore we get

$$d_m(x,z)a_m(x) - d_{m-1}(x,z)a_{m-1}(x) + x^2 d_{m-2}(x,z)a_{m-2}(x)$$
$$= -\frac{x^{m-2}}{z} + x^m(\cdots).$$

Now the left-hand side must be a polynomial of degree less than m. Therefore we have in fact

$$d_m(x,z)a_m(x) - d_{m-1}(x,z)a_{m-1}(x) + x^2 d_{m-2}(x,z)a_{m-2}(x) = -\frac{x^{m-2}}{z}.$$

Now $c_m(x,z)$ satisfies the same recurrence. Since the initial values coincide, we get

Corollary 3. *For $m \geq 2$ the generating function for $a(n,m,0,z)$ is given by*

$$\sum_{n \geq 0} a(n,m,0,z)x^n = \frac{(x^{m-1}/z) + F_m(1,-x^2) + xF_{m-1}(1,-x^2)}{L_m(1,-x^2) - x^m(z + (1/z))}. \quad (18)$$

Remark. In the same way we get

$$\sum_{n \geq 0} a(n, 2m+1, m+1, z)x^n$$
$$= \frac{(1+z)x^m(F_{m+1}(1,-x^2) + xF_m(1,-x^2))}{L_{2m+1}(1,-x^2) - x^{2m+1}(z + (1/z))}.$$

For $z = -1$ the right-hand side vanishes and therefore we get again $a(n, 2m+1, m+1, -1) = 0$.

4b) For the special case $z = 1$ also simpler recurrences can be found. It is easy to verify that

$$\left(x + \frac{1}{x} - 2\right) F_m\left(x + \frac{1}{x}, -1\right)(1+x) = \frac{1}{x^m} - \frac{1}{x^{m-1}} - x^m + x^{m+1}.$$

This implies as above
$$(N - 2)F_m(N, -1)u(0) = (K^m - K^{m-1} - K^{-m} + K^{-m-1})s_{1,0}(0).$$
Therefore we get
$$(N - 2)F_m(N, -1) \sum_j K^{2jm}u(0)$$
$$= \sum_j K^{2jm}(K^m - K^{m-1} - K^{-m} + K^{-m-1})s_{1,0}(0) = 0.$$

From this we conclude as above

Theorem 4. *The sequence*
$$a(n, 2m, k, 1) = \sum_{j \in \mathbb{Z}} \left(\left\lfloor \frac{n}{\frac{n - (2m)j + k}{2}} \right\rfloor \right)$$
satisfies the recurrence relation
$$(N - 2)F_m(N, -1)a(n, 2m, k, 1) = 0. \tag{19}$$

Corollary 4. *For $m \geq 1$ the generating function for $a(n, 2m, 0, 1)$ is given by*
$$\sum_{n \geq 0} a(n, 2m, 0, 1)x^n = \frac{F_m(1, -x^2) - xF_{m-1}(1, -x^2)}{(1 - 2x)F_m(1, -x^2)}. \tag{20}$$

4c) It is again easy to verify that
$$\left(L_m\left(x + \frac{1}{x}, -1\right) - L_{m-1}\left(x + \frac{1}{x}, -1\right) \right)(1 + x)$$
$$= \frac{1}{x^m} - \frac{1}{x^{m-2}} - x^{m-1} + x^{m+1}.$$

Therefore we get
$$(L_m(K + K^{-1}, -1) - L_{m-1}(K + K^{-1}, -1)) \sum_j K^{(2m-1)j}u(0)$$
$$= \sum_j K^{(2m-1)j}(K^m - K^{m-2} - K^{-m+1} + K^{-m-1})s_{1,0}(0) = 0.$$

This implies

Theorem 5. *The sequence*
$$a(n, 2m - 1, k, 1) = \sum_{j \in \mathbb{Z}} \left(\left\lfloor \frac{n}{\frac{n - (2m - 1)j + k}{2}} \right\rfloor \right)$$

satisfies the recurrence relation

$$(L_m(N, -1) - L_{m-1}(N, -1))a(n, 2m - 1, k, 1) = 0. \qquad (21)$$

Corollary 5. *For $m \geq 2$ the generating function for $a(n, 2m - 1, 0, 1)$ is given by*

$$\sum_{n \geq 0} a(n, 2m - 1, 0, 1)x^n = \frac{L_{m-1}(1, -x^2)}{L_m(1, -x^2) - xL_{m-1}(1, -x^2)}. \qquad (22)$$

Remark. For the special cases $z = \pm 1$ numerator and denumerator of the generating function

$$\frac{(x^{m-1}/z) + F_m(1, -x^2) + xF_{m-1}(1, -x^2)}{L_m(1, -x^2) - x^m(z + (1/z))}$$

have common divisors which can be cancelled.

This can be verified by using the following identities, which are easily deduced from the representations (6) and (13) (cf. e.g. [3]):

$$L_{2m}(x, -1) - 2 = (x^2 - 4)(F_m(x, -1))^2,$$

$$L_{2m-1}(x, -1) - 2 = \frac{(L_m(x, -1) - L_{m-1}(x, -1))^2}{x - 2},$$

$$L_{2m}(x, -1) + 2 = (L_m(x, -1))^2,$$

$$L_{2m-1}(x, -1) + 2 = (x + 2)(F_m(x, -1) - F_{m-1}(x, -1))^2.$$

References

[1] ANDREWS, G. E. (1969) Some formulae for the Fibonacci sequence with generalizations. Fibonacci Quart. **7**: 113–130

[2] BRIETZKE, E. H. M. (2006) Generalization of an identity of Andrews. Fibonacci Quart. **44**: 166–171

[3] CIGLER, J. (2001) Recurrences for some sequences of binomial sums. Sitzungsber. Österr. Akad. Wiss. Wien, Math.-nat. Kl. Abt. II **210**: 61–83; http://hw.oeaw.ac.at/sitzungsberichte_und_anzeiger_collection

[4] CIGLER, J. (2004) A class of Rogers-Ramanujan type recursions. Sitzungsber. Österr. Akad. Wiss. Wien, Math.-nat. Kl. Abt. II **213**: 71–93; http://hw.oeaw.ac.at/sitzungsberichte_und_anzeiger_collection

[5] CIGLER, J. (2005) Fibonacci-Zahlen, Gitterpunktwege und die Identitäten von Rogers-Ramanujan. Math. Semesterber. **52**: 97–125

[6] SCHUR, I. (1917) Ein Beitrag zur additiven Zahlentheorie und zur Theorie der Kettenbrüche. In: SCHUR, I. (1973) Gesammelte Abhandlungen, Bd. 2, pp. 117–136. Springer, Berlin Heidelberg New York

Author's address: Prof. Dr. Johann Cigler, Fakultät für Mathematik, Universität Wien, Nordbergstraße 15, 1090 Wien, Austria. E-Mail: Johann.Cigler@univie.ac.at.

Sitzungsber. Abt. II (2006) 215: 155–171

Sitzungsberichte

Mathematisch-naturwissenschaftliche Klasse Abt. II
Mathematische, Physikalische und Technische Wissenschaften

Die Linienelemente des $\mathbb{P}^3$

Von

Boris Odehnal

(Vorgelegt in der Sitzung der math.-nat. Klasse am 16. November 2006
durch das k. M. Hellmuth Stachel)

Zusammenfassung

Die vorliegende Arbeit befasst sich mit der Abbildung der Linienelemente des $\mathbb{P}^3$ auf die Punkte einer fünfdimensionalen rationalen Fläche M^5, die in den $\mathbb{P}^9$ eingebettet ist. Ferner wird der Zusammenhang der SEGRE-Varietät $S_{3,5}$ und der Mannigfaltigkeit M^5 untersucht. Auf der Modellfläche erscheinen einfache Linienelementmannigfaltigkeiten als projektive Unterräume wieder. Das Dualitätsprinzip der projektiven Geometrie gestattet die Übertragung der Ergebnisse über Linienelemente auf ihre dualen Gegenstücke.

Mathematics Subject Classification (AMS 2000): 51A25, 51A45.
Key words: Linienelement, SEGRE-Varietät, GRASSMANN-Varietät, Punktmodell, algebraische Fläche, rationale Fläche, Linienelementmannigfaltigkeit.

1. Einleitung

Die k-dimensionalen projektiven Unterräume eines n-dimensionalen projektiven Raumes $\mathbb{P}^n$ können bijektiv auf die Punkte einer algebraischen Varietät $G_{n,k}$ abgebildet werden [5, 6]. Die Varietät $G_{n,k}$ heißt GRASSMANN-Varietät und spannt einen projektiven Raum $\mathbb{P}^m$ von $m = \binom{n+1}{k+1} - 1$ Dimensionen auf. Die in ihr enthaltenen projektiven Unterräume entsprechen den linearen Mannigfaltigkeiten k-dimensionaler projektiver Unterräume. Die projektiven Kollineationen des $\mathbb{P}^n$ induzieren im Raum $\mathbb{P}^m$ projektive Kollineationen, die $G_{n,k}$ als Ganzes, aber nicht punktweise festhalten.

Naheliegend scheint nun die Verallgemeinerung der Geometrie k-dimensionaler Unterräume des $\mathbb{P}^n$ dahingehend, dass man etwa ein Punktmodell der Menge aller geordneten s-Tupel $(\mathbb{P}_1^{k_1}, \dots, \mathbb{P}_s^{k_s})$ k_i-dimensionaler projektiver Unterräume $\mathbb{P}_i^{k_i}$ des $\mathbb{P}^n$ konstruieren möchte. Dies führt zu den so genannten SEGRE-Varietäten $S_{k_1,\dots,k_s}$, die in einem $(k_1 + 1) \cdot \dots \cdot (n_k + 1) - 1$-dimensionalen projektiven Raum enthalten sind. Die Reihenfolge der Parameterräume $\mathbb{P}_i^{k_i}$ ist willkürlich, siehe [5]. Nehmen wir nun an, es gelte $\dim \mathbb{P}_1^{k_1} = k_1 < \dots < k_s = \dim \mathbb{P}^{k_s}$, so bilden $\mathbb{P}_i^{k_i}$ eine aufsteigende Kette $\mathbb{P}_1^{k_1} \subset \dots \subset \mathbb{P}_s^{k_s}$ von Unterräumen des $\mathbb{P}^n$, und man nennt ein solches s-Tupel *verallgemeinertes Raumelement* oder *Flagge*. Insbesondere spricht man von einer *vollständigen Flagge*, wenn alle natürlichen Zahlen aus der Menge $\{0, \dots, n-1\}$ als Dimensionszahlen auftreten.

Die Geometrie der verallgemeinerten Raumelemente sowie ihr Zusammenhang mit der Darstellungstheorie der endlichen Gruppen war bereits Gegenstand zahlreicher Untersuchungen, siehe hierzu etwa [1–4, 7].

Da sich in letzter Zeit Anwendungen für Geometrien dieser Art, insbesondere für die Geometrie der Linienelemente des euklidischen Dreiraumes, etwa in der Flächenerkennung und -rekonstruktion fanden [8], war die Geometrie der Linienelemente des euklidischen $\mathbb{R}^3$ Gegenstand neuerer Untersuchungen [10]. Auch die Geometrie der Flaggen des euklidischen Dreiraumes wurde erst kürzlich beleuchtet und in Verbindung zur Kinematik euklidischer und zur Geometrie nichteuklidischer Räume gebracht [9].

Im Folgenden soll ein Punktmodell für die Menge der Linienelemente des projektiven Raumes $\mathbb{P}^3$ über einem beliebigen kommutativen Körper $\mathbb{K}$ der Charakteristik 0 konstruiert werden, das im Gegensatz zur Segre-Varietät $S_{3,5}$ in einen neundimensionalen projektiven Raum eingebettet ist. Dazu werden zunächst in Abschnitt 2 grundlegende Tatsachen aus der analytischen und der projektiven Geometrie sowie der Liniengeometrie aufbereitet. Im darauffolgenden Abschnitt 3 werden Koordinaten für Linienelemente des $\mathbb{P}^3$ erklärt. Linienelemente des $\mathbb{P}^3$ werden auf Punkte einer fünfdimensionalen Fläche $M^5 \subset \mathbb{P}^9$ abgebildet. Diese gestattet rationale Parametrisierungen und ist vom Grad 5. Anschließend werden in Abschnitt 4 die Bilder einfacher Linienelementmannigfaltigkeiten bestimmt. Es sind diese im Wesentlichen Geraden oder Ebenen, von denen jedoch bestimmte Punkte auszunehmen sind. Das Dualitätsprinzip der projektiven Geometrie gestattet in Abschnitt 5, die Ergebnisse der Untersuchungen über die Linienelemente auf ihre dualen Gegenstücke, die Paare bestehend aus einer Ebene und einer in ihr enthaltenen Geraden, zu übertragen. Das dabei entstehende Punktmodell unterscheidet sich von Ersterem weder in den geometrischen noch in den algebraischen Eigenschaften.

2. Linienelemente des $\mathbb{P}^3$

2.1. Grundlagen, Koordinaten für Punkte, Geraden und Ebenen

Um unserem Ziel, der Koordinatisierung der Linienelemente des $\mathbb{P}^3$, näherzukommen, beschreiben wir Punkte X des durch projektiven Abschluss erzeugten $\mathbb{P}^3$ durch ihre homogenen Koordinaten $X = (x, x_0)\mathbb{K} = (x_1, x_2, x_3; x_0)\mathbb{K}$, wobei $\mathbb{K}$ ein beliebiger kommutativer Körper der Charakteristik 0 sein soll. Die vierte, homogenisierende Koordinate wird in dieser Arbeit absichtlich an die letzte Stelle gesetzt. Dies vereinfacht später die Notation.

Die Ebenen des $\mathbb{P}^3$ bilden gleichfalls einen projektiven Dreiraum, den wir mit $\mathbb{P}^{3\star}$ bezeichnen und Dualraum nennen wollen. Eine Ebene E werde durch ihren homogenen Koordinatenvektor $E = (e, e_0)\mathbb{K} = (e_1, e_2, e_3; e_0)\mathbb{K}$ beschrieben.

Einer Geraden G des $\mathbb{P}^3$, die Verbindung zweier Punkte X und Y ist, ordnen wir homogene PLÜCKER-Koordinaten $(g_1, g_2, g_3; g_4, g_5, g_6)\mathbb{K}$ zu, wie dies etwa in [11, 13] beschrieben ist. Wir fassen die Koordinaten der Punkte X und Y in der Matrix

$$\begin{bmatrix} x_0 & x_1 & x_2 & x_3 \\ y_0 & y_1 & y_2 & y_3 \end{bmatrix} \tag{1}$$

zusammen und berechnen der Reihe nach die Determinanten der Untermatrizen mit den Spaltenindizes $[(0, 1), (0, 2), (0, 3), (2, 3), (3, 1), (1, 2)]$.

Die PLÜCKER-Koordinaten einer Geraden G sind zum einen unabhängig von der Wahl der Punkte $X \in G$ und $Y \in G$ und zum anderen homogen. Letzteres ermöglicht die Interpretation dieser Sextupel als homogene Koordinaten von Punkten eines fünfdimensionalen projektiven Raumes $\mathbb{P}^5$.

Die Abbildung $\gamma \colon G \to (g_1, \ldots, g_6)\mathbb{K}$ heißt KLEINsche Abbildung. Sie ist nicht surjektiv. Fasst man die Koordinaten von G zu Vektoren $g = (g_1, g_2, g_3)$ und $\bar{g} = (g_4, g_5, g_6)$ zusammen, dann überzeugt man sich leicht, dass

$$\langle g, \bar{g} \rangle = g_1 g_4 + g_2 g_5 + g_3 g_6 = 0 \tag{2}$$

gilt. Hier und im Folgenden soll $\langle x, y \rangle = x_1 y_1 + x_2 y_2 + x_3 y_3$ gelten. In Gl. (2) ist 0 als $0_\mathbb{K}$ zu lesen. Im Folgenden wird der besseren Lesbarkeit wegen auf den verzierenden Index $_\mathbb{K}$ verzichtet. Auch der Nullvektor des $\mathbb{K}^n$ wird nur als 0 geschrieben.

Offensichtlich werden von γ nur jene Punkte des $\mathbb{P}^5$ getroffen, die auf der durch (2) beschriebenen Quadrik M_2^4 liegen. Diese Quadrik ist die GRASSMANN-Mannigfaltigkeit $G_{3,1}$ [5] und wird gelegentlich auch

PLÜCKER- oder KLEIN-Quadrik genannt. Umgekehrt ist jedes nicht-triviale Sextupel $(g_1, \ldots, g_6) \in \mathbb{K}^6$, das (2) erfüllt, ein Koordinaten-vektor genau einer Geraden G des $\mathbb{P}^3$.

2.2. Die Segre-Varietät $S_{3,5}$ und Linienelemente in $\mathbb{P}^3$

Wir benötigen folgenden Begriff:

Definition 2.1. Ein Linienelement ist ein Paar (P, G), bestehend aus einem Punkt P und einer Geraden G mit $P \in G$.

Um zunächst die Gesamtheit aller Paare, bestehend aus Punkt und Gerade des $\mathbb{P}^3$, zu beschreiben, folgen wir [5]. Wir identifizieren die Punkte des $\mathbb{P}^3$ beziehungsweise des $\mathbb{P}^5$ mit den eindimensionalen Unterräumen des $\mathbb{K}^4$ beziehungsweise $\mathbb{K}^6$ und bilden das Tensor-produkt $\mathbb{K}^4 \otimes \mathbb{K}^6 =: V$.

Sind $(p_1, p_2, p_3; p_0)$ Koordinaten in $\mathbb{K}^4$ und sind $(g_1, \ldots, g_6)$ Koordinaten in $\mathbb{K}^6$, dann erklären wir

$$w_{ij} := p_i g_j, \quad i \in \{0, 1, 2, 3\}, \quad j \in \{1, 2, 3, 4, 5, 6\} \quad (3)$$

als Koordinaten in V. Man sieht sofort, dass dim $V = 24$ gilt. Da nun sowohl die Punkt- als auch die Geradenkoordinaten homogen sind, sind auch die Koordinaten w_{ij} homogen. Sie können folglich als ho-mogene Punktkoordinaten in einem projektiven Raum $\mathbb{P}^{23}$ gedeutet werden.

Wohl gibt es zu allen Paaren (P, G) gemäß Konstruktion einen entsprechenden Punkt in $\mathbb{P}^{23}$, die Umkehrung aber gilt nicht. Bei der Tensorproduktbildung wurden alle Vektoren des $\mathbb{K}^6$ verwendet, auch jene, die nicht zu Punkten von M_2^4 zeigen. Diese Punkte entsprechen den Gewinden, siehe [11], womit in V auch die Paare bestehend aus Punkten und Gewinden enthalten sind. Um diese auszuschließen, sind den Koordinaten w_{ij} die aus (2) folgenden Bedingungen

$$w_{i_1 1} w_{j_1 4} + w_{i_2 2} w_{j_2 5} + w_{i_3 3} w_{j_3 6} = 0, \quad (4)$$

aufzuerlegen. Dabei sind die Indizes i_1, i_2, i_3, j_1, j_2 und j_3 so mit Werten aus $\{1, 2, 3\}$ zu belegen, dass nach Einsetzen der Definitions-gleichungen (3) der Koordinaten w_{ij} in (4) sich stets der Faktor $p_i p_j$ oder eben (2) abspaltet. Es gibt, wie man nach kurzer Überlegung feststellen kann, 128 Gleichungen vom Typ (4), die aber nicht völlig unabhängig sind.

Die siebendimensionale Untermannigfaltigkeit des $\mathbb{P}^{23}$, die durch (4) beschrieben wird, heißt SEGRE-Mannigfaltigkeit $S_{3,5}$. Sie ist ein Punktmodell für die Paare von Punkten und Geraden des $\mathbb{P}^3$.

Die Mannigfaltigkeit $\mathcal{L}$ der Linienelemente ist eine fünfdimensionale Teilmannigfaltigkeit der $S_{3,5}$, da ihr nur jene Paare (P, G) von Punkten und Geraden angehören, für die $P \in G$ gilt. Mit Hilfe der Inzidenzbedingungen für Punkte und Geraden gelingt es, diese Mannigfaltigkeit einzugrenzen. Ist $P = (p, p_0)\mathbb{K}$ ein Punkt auf $G = (g, \bar{g})\mathbb{K}$, so gilt (wie etwa in [11] nachzulesen ist)

$$\langle p, \bar{g} \rangle = 0 \qquad \text{und} \qquad -p_0 \bar{g} + p \times g = 0, \qquad (5)$$

wobei $p \times g$ das Kreuzprodukt der Vektoren p und g bezeichnet. Die vier linearen Gleichungen (5) können mit (3) in der Form

$$w_{14} + w_{25} + w_{36} = 0,$$

$$-w_{04} + w_{23} - w_{32} = 0,$$

$$-w_{05} + w_{31} - w_{13} = 0,$$

$$-w_{06} + w_{12} - w_{21} = 0 \qquad (6)$$

angeschrieben werden. Es gilt daher:

Satz 2.1. *Die Mannigfaltigkeit $\mathcal{L}$ der Linienelemente des $\mathbb{P}^3$ ist im Schnitt der SEGRE-Varietät $S_{3,5}$ mit einem $\mathbb{P}^{19}$ enthalten.*

Beweis. Die Gln. (6) beschreiben jede für sich eine Hyperebene des $\mathbb{P}^{23}$. Der Durchschnitt dieser vier Hyperebenen ist ein $\mathbb{P}^{19}$, wie man leicht nachrechnet. $\qquad\square$

Es ist bekannt, dass eine n-dimensionale Mannigfaltigkeit M mit höchstens abzählbar vielen Zusammenhangskomponenten stets in einen $\mathbb{R}^{2n}$ eingebettet werden kann, siehe [14].

Wir wollen im nächsten Abschnitt zeigen, dass es möglich ist, die Linienelementmannigfaltigkeit $\mathcal{L}$ in einen projektiven Raum $\mathbb{P}^9$ von neun Dimensionen so einzubetten, dass in ihr enthaltene projektive Unterräume, die linearen Mannigfaltigkeiten von Linienelementen entsprechen, weitestgehend erhalten bleiben. Darüber hinaus wird es mit der hier gezeigten Koordinatisierung der Linienelementmannigfaltigkeit möglich sein, Paare von Linienelementen zu kennzeichnen.

Das dabei entstehende Punktmodell wird nicht wie im obigen Falle durch Tensorproduktbildung der beteiligten Vektorräume, sondern durch Summenbildung erzeugt: Wir betten $\mathbb{K}^6 \setminus 0 \cong \mathbb{P}^5$ in $\mathbb{K}^{10}$ als Unterraum $x_7 = x_8 = x_9 = x_{10} = 0$ ein. $\mathbb{K}^4 \setminus 0 \cong \mathbb{P}^3$ betten wir in den Komplementärraum $x_1 = \cdots = x_6 = 0$ ein. Folglich gilt $\mathbb{K}^6 \oplus \mathbb{K}^4 = \mathbb{K}^{10}$ oder eben $\mathbb{P}^5 \vee \mathbb{P}^3 = \mathbb{P}^9$.

3. Ein anderes Modell

Es seien $(g, \overline{g})\mathbb{K}$ homogene PLÜCKER-Koordinaten einer Geraden G in $\mathbb{P}^3$ und ferner $(\widehat{g}, g_0)\mathbb{K}$ homogene Koordinaten eines Punktes P in $\mathbb{P}^3$. P liegt genau dann auf G, wenn

$$\langle \widehat{g}, \overline{g} \rangle = 0 \qquad \text{und} \qquad -g_0\overline{g} + \widehat{g} \times g = 0 \tag{7}$$

gilt. Wir erklären nun Koordinaten für Linienelemente des $\mathbb{P}^3$ auf folgende Weise:

Definition 3.1. Der Vektor $(g, \overline{g}, \widehat{g}, g_0) \in \mathbb{K}^{10}$ ist Koordinatenvektor eines Linienelements (P, G) des $\mathbb{P}^3$, wobei $(g, \overline{g})\mathbb{K}$ homogene PLÜCKER-Koordinaten der Geraden $G \subset \mathbb{P}^3$ und $(\widehat{g}, g_0)\mathbb{K} \in \mathbb{P}^3$ homogene Koordinaten des Punktes P sind und Gln. (2) und (7) gelten. Auszuschließen sind jene Vektoren, für die $g = \overline{g} = 0$ oder $\widehat{g} = 0$ und $g_0 = 0$ gilt.

Bemerkung 3.1. Die Punkte des vierdimensionalen Unterraumes $g = \overline{g} = 0 \subset \mathbb{K}^{10}$ und die Punkte des sechsdimensionalen Unterraumes $\widehat{g} = 0$, $g_0 = 0$ entsprechen keinem Linienelement, da $g = \overline{g} = 0$ keine Gerade und $\widehat{g} = 0$, g_0 keinen Punkt beschreiben.

Da die Koordinaten der Punkte und der Geraden homogen sind, sind nach Konstruktion auch die Koordinaten der Linienelemente homogen. Sie können daher als Punktkoordinaten in einem $\mathbb{P}^9$ interpretiert werden. Wir bezeichnen mit $B_1, \ldots, B_9$ und B_0 die zehn Basispunkte. Sie entsprechen im analytischen Modell den kanonischen Basisvektoren des $\mathbb{K}^{10}$.

Bemerkung 3.2. Die in Bemerkung 3.1 genannten Unterräume des Vektorraumes $\mathbb{K}^{10}$ bestimmen komplementäre projektive Unterräume $\mathbb{A}^3$ und $\mathbb{A}^5$ des $\mathbb{P}^9$.

Bemerkung 3.3. Die in Definition 3.1 erklärten Koordinaten für Linienelemente des $\mathbb{P}^3$ sind auf merkwürdige Weise homogen: Zum einen verändert das Multiplizieren des Vektors $(g, \overline{g}, \widehat{g}, g_0)$ mit einem Faktor $\lambda \in \mathbb{K} \setminus \{0\}$ nichts am geometrischen Objekt. Zum anderen kann man die beiden Bestandteile $(g_0, \widehat{g})\mathbb{K}$ und $(g, \overline{g})\mathbb{K}$ mit unterschiedlichen Faktoren $\lambda, \mu \in \mathbb{K}^2 \setminus \{0, 0\}$ multiplizieren und erhält mit $(\lambda g, \lambda \overline{g}, \mu \widehat{g}, \mu g_0)$ wieder Koordinaten, die dasselbe geometrische Objekt beschreiben. Die Inzidenzbedingungen (7) und die Bedingung (2) an die PLÜCKER-Koordinaten von G bleiben erfüllt.

Die Abbildung $(P, G) \mapsto (g, \overline{g}, \widehat{g}, g_0)\mathbb{K}$ der Linienelemente des $\mathbb{P}^3$ auf die Punkte des $\mathbb{P}^9$ ist nicht surjektiv. Die Mannigfaltigkeit M^5

der Linienelemente ist durch die Gln. (2) und (7) mit den in Definition 3.1 genannten Ausnahmen beschrieben. Wir können folgendes Resultat beweisen:

Satz 3.1. *Die Linienelemente des* $\mathbb{P}^3$ *können auf Punkte einer fünfdimensionalen algebraischen Fläche* M^5 *abgebildet werden. Die Fläche* M^5 *ist im Durchschnitt der durch die Gln. (2) und (7) beschriebenen Mannigfaltigkeit enthalten, ist vom Grad 5 und gestattet eine rationale Parametrisierung.*

Beweis. Die Dimension von M^5 ist leicht zu klären: Vier Freiheitsgrade besitzt die Geradenkomponente G des Linienelements, ein weiterer steht für den Punkt P auf G zur Verfügung.

Wir konstruieren nun eine rationale Parametrisierung: Es sei zunächst $g = (2x_1, 2x_2, 1 - x_1^2 - x_2^2)N^{-1}$ mit $N = 1 + x_1^2 + x_2^2$. Nun gilt neben $\langle g, g \rangle = 1$ auch $\|g_{,1}\| = \|g_{,2}\| = 2N^{-1}$ und $\langle g_{,1}, g_{,2} \rangle = 0$, wobei vereinfachend $_{,i} = \partial/\partial x_i$ steht. Wir verwenden abkürzend $c_3 = (1 - x_3^2)(1 + x_3^2)^{-1}$ und $s_3 = 2x_3(1 + x_3^2)^{-1}$. Den Momentenvektor von G können wir damit durch

$$\overline{g} = x_4 \left(\frac{c_3}{\|g_{,1}\|} g_{,1} + \frac{s_3}{\|g_{,2}\|} g_{,2} \right)$$

angeben. Für die Punktkomponente $(\widehat{g}, g_0)$ setzen wir zunächst $g_0 = 1$ und erhalten

$$\widehat{g} = g \times \overline{g} + \gamma g = x_4 \left(\frac{c_3}{\|g_{,2}\|} g_{,2} - \frac{s_3}{\|g_{,1}\|} g_{,1} \right) + x_5 g.$$

Wir führen mit $x_i = u_i u_0^{-1}$ ($i \in \{1, 2, 3, 4, 5\}$) homogene Parameter ein und erhalten nach anschließender Erweiterung mit dem gemeinsamen Nenner $u_0(u_0^2 + u_3^2)(u_0^2 + u_1^2 + u_2^2)$ die rationale Parametrisierung $\mathcal{M}$: $\mathbb{P}(\mathbb{K}^6) \to \mathbb{P}(\mathbb{K}^{10})$ als

$$\mathcal{M}(u_0 : u_1 : u_2 : u_3 : u_4 : u_5)$$

$$= (2u_0^2 u_1(u_0^2 + u_3^2), 2u_0^2 u_2(u_0^2 + u_3^2), u_0(u_0^2 + u_3^2)(u_0^2 - u_1^2 - u_2^2);$$

$$u_4(u_0^2 - u_3^2)(u_0^2 - u_1^2 + u_2^2) - 4u_0 u_1 u_2 u_3 u_4, -2u_1 u_2 u_4(u_0^2 - u_3^2)$$

$$+ 2u_0 u_3 u_4(u_0^2 + u_1^2 - u_2^2), -2u_0 u_1 u_4(u_0^2 - u_3^2) - 4u_0^2 u_2 u_3 u_4;$$

$$- 2u_1 u_2 u_4(u_0^2 - u_3^2) - 2u_0 u_3 u_4(u_0^2 - u_1^2 + u_2^2) + 2u_0 u_1 u_5(u_0^2 + u_3^2),$$

$$u_4(u_0^2 - u_3^2)(u_0^2 + u_1^2 - u_2^2) + 4u_0 u_1 u_2 u_3 u_4 + 2u_0 u_2 u_5(u_0^2 + u_3^2),$$

$$- 2u_0 u_2 u_4(u_0^2 - u_3^2) + 4u_0^2 u_1 u_3 u_4 + u_5(u_0^2 + u_3^2)(u_0^2 - u_1^2 - u_2^2);$$

$$u_0(u_0^2 + u_3^2)(u_0^2 + u_1^2 + u_2^2))\mathbb{K}. \tag{8}$$

Der algebraische Grad von M^5 wird ausgehend von den fünf quadratischen Formen (2) und (7) mit Hilfe des HILBERT-Polynoms [12] bestimmt. (Computeralgebrasysteme, wie zum Beispiel Maple, stellen Algorithmen zur Berechnung des HILBERT-Polynoms zur Verfügung.) Wir erhalten

$$p(t) = \frac{1}{144}t^6 + \frac{5}{48}t^5 + \frac{91}{144}t^4 + \frac{95}{48}t^3 + \frac{121}{36}t^2 + \frac{35}{12}t + 1. \quad (9)$$

Ist nun $p(t) = c_n t^n + \sum_{k=0}^{n-1} a_k t^k$ das HILBERT-Polynom einer algebraischen Varietät V, dann gilt deg $V = n!c_n$, was also im gegenständlichen Fall zu deg $M^5 = 5$ führt.

Dass die Vorgabe eines Vektors $(g, \bar{g}, \hat{g}, g_0) \in \mathbb{K}^{10}$, der sowohl (2) als auch (7) erfüllt, zwingend zu einem Linienelement führt, ist sofort einzusehen: $(g, \bar{g})\mathbb{K}$ gehören in diesem Fall wegen (2) zu genau einer Geraden des $\mathbb{P}^3$ und $(g_0, \hat{g})\mathbb{K}$ beschreibt genau einen Punkt des $\mathbb{P}^3$, soferne $(g, \bar{g})\mathbb{K} \neq (0,0)$ und $(g_0, \hat{g})\mathbb{K} \neq (0,0)$ gilt. Die Gln. (7) garantieren $P \in G$.

Punkte, die in $\mathbb{A}^5$ beziehungsweise in $\mathbb{A}^3$ liegen, sind natürlich auszunehmen. Erstere gehören einer zur PLÜCKER-Quadrik kollinearen Quadrik an, letztere füllen den ganzen $\mathbb{A}^3$, der M^5 zur Gänze angehört. Er ist der Spitzenraum eines quadratischen Kegels durch M^5 (siehe weiter unten).   □

Die angegebene rationale Darstellung ist naturgemäß nicht die einzige. Jede rationale isotherme Parametrisierung der euklidischen Einheitskugel kann auf die gezeigte Weise zur Konstruktion einer rationalen Parametrisierung von M^5 verwendet werden.

Bemerkung 3.4. Die Fläche M^5 liegt im Durchschnitt von fünf quadratischen Kegeln. Der algebraische Grad müsste nach dem Satz von BEZOUT $2^5 = 32$ sein.

Bezeichnet man die quadratischen Kegel, die durch die Gln. (2) und (7) beschrieben werden, der Reihe nach mit $\Gamma_1, \ldots, \Gamma_5$, dann sind ihre dreidimensionalen Spitzenräume $S_1 = [B_0, B_7, B_8, B_9]$, $S_2 = [B_0, B_1, B_2, B_3]$, $S_3 = [B_1, B_5, B_6, B_7]$, $S_4 = [B_2, B_4, B_6, B_8]$ und $S_5 = [B_3, B_4, B_5, B_9]$, wobei $[B_{i_1}, \ldots, B_{i_k}]$ den von den Punkten $B_{i_1}, \ldots, B_{i_k}$ aufgespannten projektiven Unterraum von $\mathbb{P}^9$ bezeichnet.

Je zwei verschiedene Spitzenräume S_i und S_j haben genau einen Punkt gemeinsam. Abb. 1 ist als Inzidenztabelle zu sehen, die die Gemeinsamkeiten und Unterschiede der Spitzenräume der Kegel Γ_i darstellt. Es handelt sich dabei um einen Zentralriss aus dem $\mathbb{P}^9$ in den $\mathbb{P}^2$.

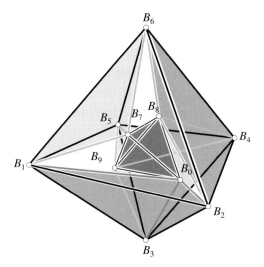

Abb. 1. Die Anordnung der Spitzenräume der quadratischen Kegel Γ_i: Die Ebenen $[B_1, B_2, B_3]$, $[B_2, B_4, B_6]$, $[B_3, B_4, B_5]$ und $[B_1, B_5, B_6]$ liegen in A^5. Der Spitzenraum $S_1 = [B_7, B_8, B_9, B_0]$ ist identisch mit $\mathbb{A}^3$

3.1. M^5 als projektives Bild der $S_{3,5}$

Wir zeigen nun, welcher Zusammenhang zwischen M^5 und $S_{3,5}$ besteht:

Satz 3.2. *Es gibt mindestens eine Projektion (singuläre Kollineation) $\mathbb{P}^{23} \to \mathbb{P}^9$, die das in der* SEGRE-*Mannigfaltigkeit enthaltene Modell der Linienelemente des $\mathbb{P}^3$ in das in Abschnitt 3 beschriebene Modell überführt.*

Beweis. Die Projektion $\pi\colon \mathbb{P}^{23} \to \mathbb{P}^9$ sei durch ihre Abbildungsgleichungen

$$\pi(w_{01}, \ldots, w_{36})\mathbb{K} = (w_{01}, w_{02}, w_{03}, w_{04}, w_{05}, w_{06}; w_{11}, w_{21}, w_{31}; w_{01})\mathbb{K} \tag{10}$$

gegeben. Die linke Seite von (10) wird nun mit (3) vereinfacht, und es ergibt sich für den Bildpunkt

$$\left(w_{01}, w_{02}, w_{03}, w_{04}, w_{05}, w_{06}; w_{11}, w_{21}, w_{31}; w_{01}\right)\mathbb{K}$$
$$= (p_0 g_1, p_0 g_2, p_0 g_3, p_0 g_4, p_0 g_5, p_0 g_6; g_1 p_1, g_1 p_2, g_1 p_3; g_1 p_0)\mathbb{K}$$
$$= (p_0 g, p_0 \overline{g}; g_1 \widehat{g}, g_1 p_0)\mathbb{K} = (g, \overline{g}, \widehat{g}, g_0)\mathbb{K},$$

wobei im letzten Schritt von der eigenartigen Homogenität der in Definition 3.1 erklärten Koordinaten für Linienelemente Gebrauch gemacht wurde. Die Projektion ist keineswegs die einzige Abbildung $S_{3,5} \to M^5$. □

3.2. Die projektiven Transformationen des $\mathbb{P}^3$

Die projektiven Kollineationen des $\mathbb{P}^3$ induzieren projektive Kollineationen des Modellraumes. Eine Kollineation $\kappa: \mathbb{P}^3 \to \mathbb{P}^3$ transformiert Punkte X gemäß

$$X'\mathbb{K} = TX\mathbb{K}, \tag{11}$$

wobei $T \in \mathrm{GL}(\mathbb{K}^4)$ gilt. Wir wollen hier nur reguläre Kollineationen betrachten. Für die von der Gruppe $\mathrm{PGL}(\mathbb{P}^3)$ induzierten Kollineationen des Modellraumes gilt nun:

Lemma 3.1. *Die projektiven Kollineationen des $\mathbb{P}^3$ aus* (11) *induzieren automorphe Kollineationen der Mannigfaltigkeit M^5. Dabei gilt für die Transformation der Linienelementkoordinaten*

$$\begin{bmatrix} g' \\ \overline{g}' \\ \widehat{g}' \\ \gamma' \end{bmatrix} = \begin{bmatrix} T \wedge T & 0 \\ 0^T & T \end{bmatrix} \begin{bmatrix} g \\ \overline{g} \\ \widehat{g} \\ \gamma \end{bmatrix}, \tag{12}$$

wobei $T \wedge T$ die Matrix der durch κ im KLEIN*schen Bildraum induzierten Kollineation ist und 0 die Nullmatrix aus $\mathbb{K}^{6 \times 4}$ bezeichnet.*

Beweis. Das Bild des Punktes $P = (\widehat{g}, g_0)\mathbb{K}$ unter der durch (11) beschriebenen Kollineation des $\mathbb{P}^3$ ist der Punkt $TP\mathbb{K}$. Die von (11) induzierte automorphe Kollineation der PLÜCKER-Quadrik ist durch $G' = T \wedge TG$ gegeben. Die Blockmatrixschreibweise liefert (12). □

4. Lineare Mannigfaltigkeiten von Linienelementen und ihre Bilder

In diesem Abschnitt betrachten wir lineare Mannigfaltigkeiten von Linienelementen und deren Bilder auf der Fläche M^5. Wir beschränken uns darauf, die Bilder konkreter Linienelementmengen zu untersuchen, da diese durch projektive Kollineationen des $\mathbb{P}^3$ in jede andere Linienelementmenge gleichen Typs übergeführt werden können.

4.1. Punktreihe

Die Punktreihe ist eine 1-parametrige Mannigfaltigkeit von Linienelementen, deren Punkte auf einer festen Geraden G variieren.

Abb. 2. Einfache Linienelementmannigfaltigkeiten: 1. Punktreihe, 2. Geradenbüschel, 3. Geradenbündel

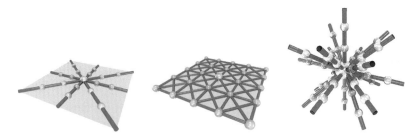

Abb. 3. Einfache Linienelementmannigfaltigkeiten: 4. Linienelementenbüschel, 5.
Linienelementenfeld, 6. Linienelementenbündel

Es sei $G = (1, 0, 0; 0, 0, 0)\mathbb{K}$ und ferner $P = (t_1/t_0, 0, 0; 1)\mathbb{K}$.
Dann sind alle Linienelemente der Punktreihe durch $B(t_0, t_1) =$
$(t_0, 0, 0; 0, 0, 0; t_1, 0, 0; t_0)\mathbb{K}$ beschrieben. Offensichtlich handelt es
sich hierbei um eine Gerade in M^5. Von dieser ist der zu $t_0 = 0$
gehörende Schnittpunkt B_7 mit $\mathbb{A}^3$ auszunehmen. Zu $\mathbb{A}^5$ liegt diese
Gerade windschief.

4.2. Geradenbüschel

Das Geradenbüschel ist ebenfalls eine 1-parametrige Mannigfaltigkeit von Linienelementen. Die Punktkomponente P ist fix. Die durch
P gehenden Geraden variieren innerhalb einer festen Ebene durch P.
Ohne Einschränkung der Allgemeinheit können wir annehmen,
dass $P = (0, 0, 0; 1)\mathbb{K}$ gilt und $E = (0, 0, 1; 0)\mathbb{K}$ die Ebene ist. Die
Geraden des Büschels seien durch $G = (1, t_1/t_0, 0; 0, 0, 0)\mathbb{K}$ gegeben,
und wir haben dann insgesamt $B(t_0, t_1) = (t_0, t_1, 0; 0, 0, 0; 0, 0, 0; t_0)\mathbb{K}$.
Auch hierbei handelt es sich um eine Gerade in M^5. Der Schnittpunkt
mit $\mathbb{A}^5$ ist B_2, er entspricht keinem Linienelement. Mit $\mathbb{A}^3$ existiert
kein Schnittpunkt.

4.3. Geradenbündel

Das Geradenbündel ist eine 2-parametrige Mannigfaltigkeit von Linienelementen, die die Punktkomponente P teilen. Die Geraden durchlaufen das Bündel um P.

Wir nehmen daher $P = (0,0,0;1)\mathbb{K}$ als Punktkomponente an. Für die Geraden haben wir dann $G = (1,t_1/t_0,t_2/t_0;0,0,0)\mathbb{K}$ und damit für das Bündel die Bildmenge $B(t_0,t_1,t_2) = (t_0,t_1,t_2;0,0,0;0,0,0;t_0)\mathbb{K}$ in M^5. Von dieser Ebene des $\mathbb{P}^9$ ist die Gerade $[B_1,B_2] \subset \mathbb{A}^5$, die $t_0 = 0$ entspricht, auszuschließen. Sie trägt keine Punkte, die Linienelementen des $\mathbb{P}^3$ entsprechen. Ferner gilt $\mathbb{A}^3 \cap B(t_0,t_1,t_2) = \emptyset$.

4.4. Linienelementenbüschel

Das Linienelementenbüschel besteht aus allen Linienelementen, deren Geraden einem Büschel angehören. Es handelt sich dabei offenbar um eine zweidimensionale Mannigfaltigkeit von Linienelementen.

Mit $P = (t_1/t_0,t_2/t_0,0;1)\mathbb{K}$ und $G = (t_1/t_0,t_2/t_0,0;0,0,0)\mathbb{K}$ erhält man $B(t_0,t_1,t_2) = (t_1,t_2,0;0,0,0;t_1,t_2,0;t_0)\mathbb{K}$ als Beschreibung dieser Linienelementenfamilie. Es ist dies eine Ebene in M^5. Sie schneidet $\mathbb{A}^5$ gar nicht und $\mathbb{A}^3$ im Punkt B_0.

4.5. Linienelementenfeld

Das Linienelementenfeld besteht aus allen Linienelementen, die in einer Ebene E liegen. Es handelt sich dabei um eine 3-parametrige Mannigfaltigkeit von Linienelementen.

Es sei also $E = (0,0,1;0)\mathbb{K}$. Die Punkte sind durch $P = (t_1/t_0, t_2/t_0,0;1)\mathbb{K}$ beschrieben und die Geraden folglich durch $G = (t_0^2,t_3t_0,0;0,0,t_1t_3-t_2t_0)\mathbb{K}$, womit sich für das Feld $B(t_0,t_1,t_2,t_3) = (t_0^2,t_0t_3,0; 0,0,t_1t_3-t_0t_2;t_0t_1,t_0t_2,0;t_0^2)\mathbb{K}$ ergibt. Es handelt sich hierbei um eine ringartige Quadrik in M^5. Ihre Erzeugenden entsprechen den Punktreihen und Geradenbüschel, die im Linienelementenfeld enthalten sind. Von dieser ist einzig der Punkt $B_6 \in \mathbb{A}^5$ auszunehmen.

4.6. Linienelementenbündel

Das Linienelementenbündel ist die Menge aller Linienelemente, deren Geraden einem Bündel angehören. Die Punktkomponenten durchlaufen alle Geraden des Bündels. Es handelt sich hierbei um eine 3-parametrige Mannigfaltigkeit von Linienelementen.

Auch hier genügt es, ein spezielles Geradenbündel zu betrachten: Wir wählen das Geradenbündel um den Punkt $B_0 = (1,0,0,0)\mathbb{K} \in \mathbb{P}^3$.

Dann ist $P = (t_0, t_1, t_2, t_3)\,\mathbb{K}$, und das Linienelementenbündel ist durch $B(t_0, t_1, t_2, t_3) = (t_1, t_2, t_3; 0, 0, 0; t_1, t_2, t_3, t_0)\,\mathbb{K}$ beschrieben. Als Bild im Modellraum erscheint ein dreidimensionaler Unterraum des $\mathbb{P}^9$, vermindert um den in $\mathbb{A}^3$ gelegenen Punkt B_0. Der Durchschnitt mit $\mathbb{A}^5$ ist leer.

Zusammenfassend gilt daher:

Satz 4.1. *Die Abbildung der Linienelemente auf Punkte der $M^5 \subset \mathbb{P}^9$ bildet*

1. die Punktreihe auf eine in M^5 gelegene und um einen Punkt des $\mathbb{A}^3$ verminderte Gerade,

2. das Geradenbüschel auf eine in M^5 gelegene und um einen Punkt des $\mathbb{A}^5$ verminderte Gerade,

3. das Geradenbündel auf eine in M^5 gelegene und um eine Gerade des $\mathbb{A}^5$ verminderte Ebene,

4. das Linienelementenbüschel auf eine in M^5 gelegene und um einen Punkt des $\mathbb{A}^3$ verminderte Ebene,

5. das Linienelementenbündel auf einen in M^5 gelegenen und um einen Punkt des $\mathbb{A}^5$ verminderten dreidimensionalen projektiven Unterraum und

6. das Linienelementenfeld auf eine in M^5 gelegene und um einen Punkt des $\mathbb{A}^5$ verminderte ringartig Quadrik ab.

4.7. Kennzeichnung der Linienelementpaare

Wir betrachten Paare $((P, G), (Q, H))$ von Linienelementen mit den Koordinaten $(g, \overline{g}, \widehat{g}, g_0)$ und $(h, \overline{h}, \widehat{h}, h_0)$. Im projektiven Raum können diese Paare hinsichtlich des Schnittverhaltens beziehungsweise der Inzidenz von Punkten und Geraden unterschieden werden. Dies kann mit Hilfe der den Linienelementen zugeordneten Koordinaten auch analytisch geschehen. Dabei sind die folgenden sechs Fälle zu unterscheiden:

Fall 1. Die Geraden G und H sind windschief. Dies kann als der allgemeine Fall bezeichnet werden. Er besitzt keinerlei analytische Kennzeichnung.

Fall 2. Die Geraden G und H schneiden einander in einem von P und Q verschiedenen Punkt. Die KLEINschen Bilder von G und H liegen dann bezüglich der KLEINschen Quadrik M_2^4 polar, und es gilt

$$\langle g, \overline{h} \rangle + \langle \overline{g}, h \rangle = 0, \tag{13}$$

womit auch das Linienelementpaar gekennzeichnet ist.

Fall 3. Die Geraden G und H schneiden einander in Q. (Falls sie einander in P schneiden, ist durch Umbenennung der Objekte dieser Fall erreichbar.) Neben (13) gilt nun auch

$$\langle \widehat{h}, g \rangle = 0 \quad \text{und} \quad -h_0 g + h \times \overline{g} = 0. \quad (14)$$

Fall 4. Falls die beiden Linienelemente den Punkt teilen, also $P = Q$ gilt, dann schneiden einander auch G und H. Es gilt dann nicht nur (13), sondern auch $(\widehat{g}, g_0) = \lambda(\widehat{h}, h_0)$, wobei $\lambda \in \mathbb{K} \setminus \{0\}$ ist.

Fall 5. Stimmen die Linienelemente in ihren Geradenkomponenten überein, ohne dass dabei $P = Q$ gilt, dann gilt für die Koordinaten $(g, \overline{g}) = \lambda(h, \overline{h})$ mit einem geeigneten $\lambda \in \mathbb{K} \setminus \{0\}$.

Fall 6. Identische Linienelemente sind durch $(g, \overline{g}) = \lambda(h, \overline{h})$ und $(\widehat{g}, g_0) = \mu(\widehat{h}, h_0)$ mit $(\lambda, \mu) \in \mathbb{K}^2 \setminus \{0, 0\}$ gekennzeichnet. Dabei müssen λ und μ keineswegs gleich sein.

5. Duale Linienelemente

Die dualen Gegenstücke der Linienelemente sind Paare (G, E), bestehend aus einer Ebene E und einer in ihr gelegenen Geraden G. Man könnte sie *Ebenenelemente* nennen. Beschreiben wir nun die Ebene E durch ihre homogenen Koordinaten $(g_0, \widehat{g})\mathbb{K}$ und die Gerade G durch ihre homogenen PLÜCKER-Koordinaten $(g, \overline{g})\mathbb{K}$. G liegt genau dann in E, wenn

$$\langle \widehat{g}, g \rangle = 0, \quad -g_0 g + \widehat{g} \times \overline{g} = 0 \quad (15)$$

gilt.

Analog zu Def. 3.1 kann man Folgendes vereinbaren:

Definition 5.1. Der Vektor $(g, \overline{g}, \widehat{g}, g_0) \in \mathbb{K}^{10}$ ist der Koordinatenvektor eines Ebenenelements (G, E), wenn $(g, \overline{g})\mathbb{K}$ die homogenen Koordinaten der Geraden G und $(g_0, \widehat{g})\mathbb{K}$ die homogenen Koordinaten der Ebene E sind und die Gln. (2) und (15) gelten. Vektoren mit $\overline{g} = g = 0$ oder $\widehat{g} = 0$ und $g_0 = 0$ sind auszuschließen.

Die Interpretation der Koordinaten der Ebenenelemente als Koordinaten von Punkten eines neundimensionalen projektiven Raumes $\mathbb{P}^9$ ist naheliegend.

Bemerkung 5.1. Wie schon bei den Koordinaten für die Linienelemente ist auch hier eine ganz eigenartige Homogenität zu beobachten. Die PLÜCKER-Koordinaten von G und die Koordinaten der Ebene E können unabhängig voneinander mit beliebigen Faktoren aus $\mathbb{K} \setminus \{0\}$ multipliziert werden, ohne dass dabei das geometrische Objekt verändert wird.

Das Dualitätsprinzip offenbart uns folgenden Satz:

Satz 5.1. *Die Ebenenelemente des projektiven Dreiraumes $\mathbb{P}^3$ können auf die Punkte einer fünfdimensionalen algebraischen Fläche $M^{5\star} \subset \mathbb{P}^9$ mit den Gln. (2) und (15) abgebildet werden. $M^{5\star}$ ist rational parametrisierbar und ist vom Grad 5.*

Beweis. Die Ergebnisse aus Abschnitt 3 sind zu dualisieren, das heißt, wir deuten jetzt $(\widehat{g}, g_0)\mathbb{K}$ als Koordinaten einer Ebene anstatt eines Punktes. Die Dimension, der algebraische Grad und die rationale Parametrisierung folgen aus Satz 3.1. $\square$

Analog zu 3.1 gilt:

Satz 5.2. *Es gibt eine Projektion $\pi\colon \mathbb{P}^{23} \to \mathbb{P}^9$, die die in der SEGRE-Mannigfaltigkeit $S_{3,5}$ enthaltene Ebenenelementmannigfaltigkeit auf $M^{5\star}$ abbildet.*

Bemerkung 5.2. Die beiden SEGRE-Mannigfaltigkeiten $S_{3,5}$ und $S_{5,3}$ sind nicht wesentlich voneinander verschieden, siehe hierzu etwa [5].

Beweis. Der Nachweis der Behauptung besteht im Umdeuten der Punktkoordinaten im Beweis zu Satz 3.1 in Ebenenkoordinaten. $\square$

Bemerkung 5.3. Auch hier gibt es zwei windschiefe Unterräume $\mathbb{A}^{3\star}$ und $\mathbb{A}^{5\star}$ des $\mathbb{P}^9$, die keine Punkte enthalten können, welche Ebenenelementen entsprechen. Die Gleichungen sind in Bemerkung 3.1 angegeben.

Das Dualitätsprinzip gestattet uns auch die Formulierung eines zu Satz 4.1 analogen Satzes über die Ebenenelemente:

Satz 5.3. *Die Abbildung der Ebenenelemente auf Punkte der $M^{5\star} \subset \mathbb{P}^9$ bildet*

1. das Ebenenbüschel auf eine in $M^{5\star}$ gelegene und um einen Punkt des $\mathbb{A}^{3\star}$ verminderte Gerade,

2. das Geradenbüschel auf eine in $M^{5\star}$ gelegene und um einen Punkt des $\mathbb{A}^{5\star}$ verminderte Gerade,

3. die Ebenen durch die Geraden eines Büschels auf eine in $M^{5\star}$ gelegene und um einen Punkt des $\mathbb{A}^{3\star}$ verminderte Ebene,

4. das Geradenfeld auf eine um eine Gerade des $\mathbb{A}^{3\star}$ verminderte Ebene,

5. die Ebenen durch die Geraden eines Feldes auf einen in $M^{5\star}$ gelegenen und um einen Punkt des $\mathbb{A}^{3\star}$ verminderten dreidimensionalen projektiven Unterraum und

6. *die Ebenen durch die Geraden eines Bündels auf eine in $M^{5\star}$ gelegene und um einen Punkt des $\mathbb{A}^{3\star}$ verminderte ringartige Quadrik ab.*

Bemerkung 5.4. Eine analytische Kennzeichnung von Paaren dualer Linienelemente kann nun analog zu Abschnitt 4.7 geschehen.

6. Abschließende Betrachtungen

Eine mögliche Anwendung der Linienelementkoordinaten könnte in der projektiven Differentialgeometrie gefunden werden. Wir fassen beispielsweise eine Raumkurve C als Menge ihrer Linienelemente auf. Ist C durch eine Parametrisierung $C(t) = (c(t), c_0(t))$: $I \subset \mathbb{R} \to \mathbb{R}^3$ festgelegt, dann ist an jeder Stelle $t_0 \in I$ ihrer Tangenten die Verbindungsgerade des Kurvenpunktes $C(t_0)$ und des Ableitungspunktes $\dot{C}(t_0)$. Die 1-parametrige Mannigfaltigkeit der Linienelemente ist damit durch

$$\mathcal{C}(t) = (c_0\dot{c} - \dot{c}_0 c, c \times c, c, c_0)\mathbb{R} \qquad (16)$$

beschrieben. Dabei handelt es sich um eine ganz in M^5 enthaltene Kurve. Die Abbildung der Linienelemente des $\mathbb{P}^3$ auf die Punkte von M^5 bildet also den *Linienelementverband* einer Kurve auf eine *Kurve in M^5* ab. Für ebene Kurven ist die Bildkurve auf M^5 nach Satz 4.1 in einer dem Linienelementenfeld der Ebene entsprechenden Quadrik enthalten.

Kurven auf Regelflächen können gemeinsam mit den Erzeugenden der Regelfläche zu *Linienelementenstreifen* zusammengefasst werden. Das ist auch mit einem Vektorfeld längs einer Kurve möglich. Nach Satz 4.1 spannen die Bildkurven solcher *Streifen auf einem Zylinder oder Kegel* höchsten einen *dreidimensionalen projektiven Unterraum, der zur Gänze in M^5 liegt*, auf.

Ist eine Regelfläche $\mathcal{R}$ des projektiv abgeschlossenen Anschauungsraumes durch eine Parametrisierung der Form $R(t) = l(t) + vr(t)$ beschrieben (wobei wir nur $r(t) \neq 0$ voraussetzen), so ist $s = l + \langle \dot{l}, \dot{r} \rangle \langle \dot{r}, \dot{r} \rangle^{-1} r$ ihre Striktionslinie, wenn man annimmt, dass $\dot{r} \neq 0$ im betrachteten Intervall gilt, $\mathcal{R}$ also dort keine zylindrischen Erzeugenden besitzt. Das *Striktionsband* von $\mathcal{R}$ ist dann durch

$$\mathcal{S}(t) = (r, l \times r, \langle \dot{r}, \dot{r} \rangle l - \langle \dot{l}, \dot{r} \rangle r, \langle \dot{r}, \dot{r} \rangle)\mathbb{R}, \qquad (17)$$

die auf M^5 gelegene Bildkurve des Striktionsbandes, parametrisiert.

Wie bei den Regelflächen bereits geschehen, könnte man die Differentialgeometrie der Streifen, also der Regelfläche mitsamt einer darauf befindlichen Kurve, auf die Differentialgeometrie der Kurven in M^5 zurückführen.

Danksagung

An dieser Stelle möchte ich H. STACHEL für seine konstruktive Kritik und manchen Hinweis danken. J. WALLNER gebührt mein Dank für einen Hinweis die algebraische Geometrie betreffend. Schließlich sei H. POTTMANN gedankt, dessen Anregung, die Geometrie der Linienelemente im Euklidischen Dreiraum zu studieren, meine Aufmerksamkeit auch auf die Linienelemente des $\mathbb{P}^3$ gelenkt hat.

Literatur

[1] BURAU, W. (1954) Eine gemeinsame Verallgemeinerung aller Veroneseschen und Grassmannschen Mannigfaltigkeiten und die irreduziblen Darstellungen der projektiven Gruppen. Rend. Circ. Math. Palermo, II. Ser. **3**: 244–268

[2] BURAU, W. (1958) Zur Geometrie der verallgemeinerten Raumelemente des $\mathbb{P}^n$ und der zugehörigen J-Mannigfaltigkeiten. Abh. Math. Sem. Univ. Hamburg **22**: 141–157

[3] BURAU, W. (1967) Über die Hilbertfunktion der Grundmannigfaltigkeiten der allgemeinen projektiven Gruppe. Monatsh. Math. **71**: 97–99

[4] BURAU, W. (1977) Über die irreduziblen Darstellungen der klassischen Gruppen und die zugehörigen Grundmannigfaltigkeiten. In: ARNOLD, H. J., BENZ, W., WEFELSCHEID, H. (eds.) Beiträge zur geometrischen Algebra (Proc. Symp. Duisburg, 1976), pp. 63–71. Birkhäuser, Basel

[5] BURAU, W. (1961) Mehrdimensionale projektive und höhere Geometrie. VEB Dt. Verlag der Wissenschaften, Berlin

[6] GIERING, O. (1982) Vorlesungen über höhere Geometrie. Vieweg, Braunschweig Wiesbaden

[7] HAVLICEK, H., LIST, K., ZANELLA, C. (2002) On automorphisms of flag spaces. Linear Multilinear Algebra **50**: 241–251

[8] HOFER, M., ODEHNAL, B., POTTMANN, H., STEINER, T., WALLNER, J. (2005) 3D shape recognition and reconstruction based on line element geometry. In: Tenth IEEE International Conference on Computer Vision, Vol. 2, pp. 1532–1538. IEEE Computer Society

[9] ODEHNAL, B. (2006) Flags in Euclidean three-space. Mathematica Pannonica **17**: 29–48

[10] ODEHNAL, B., POTTMANN, H., WALLNER, J. (2006) Equiform kinematics and the geometry of line elements. Beitr. Algebra Geom. **47** (No. 2): 567–582

[11] POTTMANN, H., WALLNER, J. (2001) Computational Line Geometry. Springer, Berlin Heidelberg New York

[12] SHAFAREVICH, I. R. (1988) Algebraic Geometry. Springer, Berlin Heidelberg New York

[13] WEISS, E. A. (1937) Einführung in die Liniengeometrie und Kinematik. Teubner, Leipzig

[14] WHITNEY, H. (1944) The self-intersections of a smooth n-manifold in $2n$-space. Ann. of Math. **45**: 220–246

Anschrift des Verfassers: Mag. Dr. Boris Odehnal, Institut für Diskrete Mathematik und Geometrie, Technische Universität Wien, Wiedner Hauptstraße 8–10, 1040 Wien, Austria. E-Mail: boris@geometrie.tuwien.ac.at.

Sitzungsber. Abt. II (2006) 215: 173–176

Sitzungsberichte
Mathematisch-naturwissenschaftliche Klasse Abt. II
Mathematische, Physikalische und Technische Wissenschaften

© Österreichische Akademie der Wissenschaften 2007
Printed in Austria

An Extremum Problem for Convex Polygons

By

Gerhard Larcher and Friedrich Pillichshammer

(Vorgelegt in der Sitzung der math.-nat. Klasse am 14. Dezember 2006
durch das w. M. August Florian)

Abstract

In the Euclidean plane, consider a convex n-gon with unit perimeter. For a certain class of functions $f\colon [0, 1/2] \to \mathbb{R}_0^+$ we establish the least upper bound on the sum of the values of f over the distances of all pairs of vertices of the polygon.

Mathematics Subject Classification (2000): 52A40, 51M04, 51K05.
Key words: Convex polygon, Euclidean distance, sum of distances.

Let $f\colon [0, 1/2] \to \mathbb{R}_0^+$ be a function. For $n \geq 3$, let $x_1, \ldots, x_n$ be the (pairwise different) vertices of a convex polygon with unit perimeter in the Euclidean plane. Define

$$S_n(f) := \sum_{1 \leq i < j \leq n} f(\|x_i - x_j\|),$$

where $\|\cdot\|$ denotes the Euclidean norm.

We ask, what is the least upper bound on $S_n(f)$?

For the specific function $f(x) = x$ this question was stated as *open problem* in [1] and completely solved in [2, Theorem 1] (here even a best possible lower bound was given). Furthermore, in [2, Theorem 3] an upper bound was given (which is best possible for even n, but not for odd n) if $f(x) = x^2$.

In this short note we give the solution to this question for a certain class of functions. Our result generalizes [2, Theorem 1] and gives the answer to [2, Open Problem 2].

Theorem 1. *Let* $f\colon [0, 1/2] \to \mathbb{R}_0^+$ *be such that the function* $x \mapsto f(x)/x$ *attains its maximum in* $x = 1/2$. *Then for any* $n \geq 3$ *and any convex polygon with* n *vertices and with unit perimeter in the Euclidean plane we have*

$$S_n(f) \leq f\left(\frac{1}{2}\right)\left\lfloor\frac{n}{2}\right\rfloor\left\lceil\frac{n}{2}\right\rceil.$$

Furthermore the bound is approached arbitrarily closely by a convex polygon with vertices $x_1, \ldots, x_{\lfloor n/2 \rfloor}$ *that are arbitrarily close to the origin and* $x_{\lfloor n/2 \rfloor + 1}, \ldots, x_n$ *that are arbitrarily close to the point* $1/2$ *on the x-axis. If the function* $x \mapsto f(x)/x$ *attains its maximum if and only if* $x = 1/2$ *and if* $n \geq 4$, *then the above inequality is even strict.*

Remark 1. Note that it is not enough that only f attains its maximum in $x = 1/2$. For example consider the function $f(x) = \sqrt{x}$. If $n = 6m$ points $x_1, \ldots, x_{6m}$ are distributed evenly among the vertices of a regular triangle of edge-length $1/3$, then we have $S_{6m}(\sqrt{\cdot}) = \sum_{1 \leq i < j \leq 6m}\sqrt{\|x_i - x_j\|} = 12m^2/\sqrt{3} > 9m^2/\sqrt{2} = \sqrt{1/2}\lfloor 6m/2 \rfloor \lceil 6m/2 \rceil$, such that the bound from Theorem 1 is not valid any more.

For the proof of Theorem 1 we need the following elementary lemma.

Lemma 1. *Let* $f\colon [0, 1/2] \to \mathbb{R}_0^+$ *be such that the function* $x \mapsto f(x)/x$ *attains its maximum in* $x = 1/2$. *Let* $n \geq 3$ *and let* $a_1, \ldots, a_n$ *be the side-lengths of a plane convex n-gon with perimeter at most one, i.e.,* $\sum_{k=1}^n a_n \leq 1$. *Then we have*

$$\sum_{k=1}^n f(a_k) \leq 2f\left(\frac{1}{2}\right).$$

Proof. Trivially we have $a_k \leq 1/2$ for all $k = 1, \ldots, n$ and therefore

$$\sum_{k=1}^n f(a_k) = \sum_{k=1}^n \frac{f(a_k)}{a_k}a_k \leq 2f\left(\frac{1}{2}\right)\sum_{k=1}^n a_k \leq 2f\left(\frac{1}{2}\right). \qquad \square$$

Now we give the proof of Theorem 1.

Proof of Theorem 1. We use the ideas from [2, Proof of Theorem 1]. Let the vertices $x_1, \ldots, x_n$ of the polygon P be arranged clockwise.

Assume first that n is even. Now we consider the $\binom{n/2}{2}$ convex quadrangles

$$Q_{i,j} := \overline{x_i x_j x_{i+\frac{n}{2}} x_{j+\frac{n}{2}} x_i}$$

for all i and j satisfying $1 \leq i < j \leq \frac{n}{2}$. Let

$$u(i,j) := f(\|x_i - x_j\|) + f(\|x_j - x_{i+\frac{n}{2}}\|) + f(\|x_{i+\frac{n}{2}} - x_{j+\frac{n}{2}}\|)$$
$$+ f(\|x_{j+\frac{n}{2}} - x_i\|).$$

As $Q_{i,j}$ is convex and contained in P it follows from Lemma 1 that $u(i,j) \leq 2f(1/2)$. Trivially $\|x_i - x_{i+n/2}\| \leq \frac{1}{2}$ for all i such that $1 \leq i \leq \frac{n}{2}$ and $\|x_i - x_{i+n/2}\| < \frac{1}{2}$ for at least one such choice of i. Since also f attains its maximum in $x = 1/2$ we have

$$S_n(f) = \sum_{1 \leq i < j \leq \frac{n}{2}} u(i,j) + \sum_{i=1}^{\frac{n}{2}} f(\|x_i - x_{i+\frac{n}{2}}\|) \leq \binom{\frac{n}{2}}{2} 2f\left(\frac{1}{2}\right)$$
$$+ \frac{n}{2} f\left(\frac{1}{2}\right) = f\left(\frac{1}{2}\right) \frac{n^2}{4}.$$

Here the first equality can be easily checked by counting all the different distances occurring on the right-hand side. It is clear that each distance on the right side of the equality appears at most once, but on the other hand we sum up $4\binom{n/2}{2} + \frac{n}{2} = \binom{n}{2}$ distances, so the equality is true. Furthermore, if $f(x)/x < 2f(1/2)$ for all $x \in [0, 1/2)$, then also $f(x) < f(1/2)$ for all $x \in [0, 1/2)$ and the above inequality is strict as well.

Now let n be odd. In this case we consider the $\binom{n-1}{2}$ convex quadrangles

$$Q_{i,j} := \overline{x_i x_j x_{i+\frac{n+1}{2}} x_{j+\frac{n+1}{2}} x_i}$$

for $1 \leq i < j \leq \frac{n-1}{2}$ and the $\frac{n-1}{2}$ convex triangles

$$R_i := \overline{x_i x_{\frac{n+1}{2}} x_{i+\frac{n+1}{2}} x_i}$$

for $1 \leq i \leq \frac{n-1}{2}$. Let

$$u(i,j) := f(\|x_i - x_j\|) + f(\|x_j - x_{i+\frac{n+1}{2}}\|) + f(\|x_{i+\frac{n+1}{2}} - x_{j+\frac{n+1}{2}}\|)$$
$$+ f(\|x_{j+\frac{n+1}{2}} - x_i\|)$$

and

$$v(i) := f(\|x_i - x_{\frac{n+1}{2}}\|) + f(\|x_{\frac{n+1}{2}} - x_{i+\frac{n+1}{2}}\|) + f(\|x_{i+\frac{n+1}{2}} - x_i\|).$$

As $Q_{i,j}$ and R_i are both convex polygons and contained in P, it follows from Lemma 1 that $u(i,j) \leq 2f(1/2)$ and $v(i) \leq 2f(1/2)$. So

$$S_n(f) = \sum_{1 \leq i < j \leq \frac{n-1}{2}} u(i,j) + \sum_{i=1}^{\frac{n-1}{2}} v(i) \leq 2f\left(\frac{1}{2}\right)\left(\binom{\frac{n-1}{2}}{2} + \frac{n-1}{2}\right)$$

$$= f\left(\frac{1}{2}\right)\frac{n^2-1}{4}.$$

Here the first equality can be checked as above. If $n > 3$, then we must have $v(i) < 2f(1/2)$ for at least one i satisfying $1 \leq i \leq \frac{n-1}{2}$.

Hence the desired bound is proved in both cases.

It is easy to see that the bound is approached arbitrarily closely by the point distributions given in Theorem 1. We just mention that $\lim_{x \to 0^+} f(x) = 0$ since $0 < f(x)/x \leq 2f(1/2)$ for all $x \in (0, 1/2]$. $\square$

References

[1] AUDET, C., HANSEN, P., MESSINE, F. (2007) Extremal problems for convex polygons. J. Global Optim. (to appear)
[2] LARCHER, G., PILLICHSHAMMER, F. (2007) The sum of distances between vertices of a convex polygon with unit perimeter. Amer. Math. Monthly (to appear)

Authors' address: Prof. Dr. Gerhard Larcher and Prof. Dr. Friedrich Pillichshammer, Institut für Finanzmathematik, Universität Linz, Altenberger Straße 69, 4040 Linz, Austria. E-Mail: gerhard.larcher@jku.at, friedrich.pillichshammer@jku.at.

Österreichische Akademie der Wissenschaften
Mathematisch-naturwissenschaftliche Klasse

Anzeiger

Abteilung II

Mathematische, Physikalische und Technische Wissenschaften

142. Band
Jahrgang 2006

Wien 2007

Verlag der Österreichischen Akademie der Wissenschaften

Inhalt

Anzeiger Abt. II

Anzeiger Abt. II (2006) 142: 3–7

Anzeiger

Mathematisch-naturwissenschaftliche Klasse Abt. II
Mathematische, Physikalische und Technische Wissenschaften

Thomas' Family of Thue Equations over Imaginary Quadratic Fields, II

By

Clemens Heuberger, Attila Pethő, and Robert F. Tichy

(Vorgelegt in der Sitzung der math.-nat. Klasse am 23. März 2006
durch das k. M. Robert F. Tichy)

Abstract

We completely solve the family of relative Thue equations

$$x^3 - (t-1)x^2y - (t+2)xy^2 - y^3 = \mu,$$

where the parameter t, the root of unity μ and the solutions x and y are integers in the same imaginary quadratic number field. This is achieved using the hypergeometric method for $|t| \geq 53$ and BAKER's method combined with a computer search using continued fractions for the remaining values of t.

Let F be an irreducible form of degree at least 3 with integral coefficients and m be a nonzero integer. Then the Diophantine equation

$$F(x,y) = m$$

is called a *Thue* equation in honor of THUE [10] who proved that it has only finitely many solutions over the integers. Algorithms for solving single Thue equations over $\mathbb{Z}$ have been developed, see BILU and HANROT [1].

Starting with THOMAS [9] in 1990, several families of parametrized Thue equations (of positive discriminant) have been solved, cf. the surveys [4, 3].

In the last years, a few parametrized families of relative Thue equations where the parameter and the solutions are elements of an imaginary quadratic number field have been studied by the authors [6], by ZIEGLER [11, 12], and by JADRIJEVIĆ and ZIEGLER [7].
In [6], the parametrized family of Thue equations

$$x^3 - (t-1)x^2y - (t+2)xy^2 - y^3 = \mu, \tag{1}$$

for $x, y \in \mathbb{Z}_{\mathbb{Q}(t)}$, an imaginary quadratic integer t, a root of unity μ in $\mathbb{Z}_{\mathbb{Q}(t)}$ has been studied. This is the family that THOMAS [9] and MIGNOTTE [8] solved completely in the rational integer case. In [6], all solutions for $|t| > 3.023 \cdot 10^9$ have been found using BAKER's method. Furthermore, all solutions for $\operatorname{Re} t = -1/2$ were claimed to be listed. However, the proof of [6, Theorem 3] is incorrect (more precisely, the argument for excluding the possibility $\Lambda = 0$ in [6, Section 7] is invalid) and some solutions are missing in [6, Table 2].

By combining the hypergeometric method due to THUE and SIEGEL (for values $|t| \geq 53$) and lower bounds for linear forms in logarithms ("BAKER's method") together with a computer search (using continued fraction expansions) for $|t| < 53$, the Diophantine equation (1) can be solved *completely* for *all values of* t.

The details are discussed in [2]. The purpose of this note is to announce the corrected and complete result:

Theorem. *Let t be an integer in an imaginary quadratic number field, $t \notin \{(-1 \pm 3\sqrt{-3})/2\}$, $\mathbb{Z}_{\mathbb{Q}(t)}$ be the ring of integers of $\mathbb{Q}(t)$,*

$$F_t(X, Y) = X^3 - (t-1)X^2Y - (t+2)XY^2 - Y^3 \in \mathbb{Z}_{\mathbb{Q}(t)}[X, Y],$$

and μ be a root of unity in $\mathbb{Q}(t)$.
Then all solutions $(x, y) \in \mathbb{Z}^2_{\mathbb{Q}(t)}$ to

$$F_t(x, y) = \mu \tag{2}$$

Table 1. Solutions (if contained in $\mathbb{Q}(t)$) to (2) for all t, where $\omega_3 = (1 + \sqrt{-3})/2$

x	y	μ	x	y	μ	x	y	μ
0	1	−1	i	0	−i	$-1+\omega_3$	$1-\omega_3$	−1
−1	0	−1	0	i	i	ω_3	0	−1
1	−1	−1	−i	0	i	0	$1-\omega_3$	1
0	−1	1	i	−i	i	0	ω_3	1
−1	1	1	0	−ω_3	−1	$-\omega_3$	0	1
1	0	1	0	$-1+\omega_3$	−1	$1-\omega_3$	$-1+\omega_3$	1
0	−i	−i	$-\omega_3$	ω_3	−1	$-1+\omega_3$	0	1
−i	i	−i	$1-\omega_3$	0	−1	ω_3	$-\omega_3$	1

Table 2. Overview on sporadic solutions to (2) for specific t

t	Number of solutions	$\max\{|x|^2, |y|^2\}$
-4	6	81
-2	6	9
-1	12	81
0	12	81
1	6	9
3	6	81
$-1 \pm 2i$	24	5
$-1 \pm 3i$	24	5
$\pm 2i$	24	5
$\pm 3i$	24	5
$-1 \pm \sqrt{-2}$	6	9
$-1 \pm 2\sqrt{-2}$	6	3
$\pm \sqrt{-2}$	6	9
$\pm 2\sqrt{-2}$	6	3
$-2 \pm 2\sqrt{-3}$	12	688
$(-3 \pm 3\sqrt{-3})/2$	24	7
$-1 \pm \sqrt{-3}$	24	3
$-1 \pm 2\sqrt{-3}$	6	1
$(-1 \pm \sqrt{-3})/2$	18	27
$\pm \sqrt{-3}$	24	3
$\pm 2\sqrt{-3}$	6	1
$(1 \pm 3\sqrt{-3})/2$	24	7
$1 \pm 2\sqrt{-3}$	12	688
$-2 \pm \sqrt{-5}$	6	86
$1 \pm \sqrt{-5}$	6	86
$-1 \pm \sqrt{-7}$	12	4
$(-1 \pm \sqrt{-7})/2$	6	7
$\pm \sqrt{-7}$	12	4
$(-3 \pm \sqrt{-11})/2$	6	20
$(1 \pm \sqrt{-11})/2$	6	20
$(-1 \pm \sqrt{-19})/2$	6	19
$(-1 \pm \sqrt{-31})/2$	6	98
$(-1 \pm \sqrt{-35})/2$	6	611

are listed in Table 1 (solutions independent of t) and in the online table [5] (solutions for specific values of t). A short summary of these 732 "sporadic" solutions is given in Table 2. The sporadic solutions with $\operatorname{Re} t = -1/2$ *are listed in Table 3.*

Remark. If $t \in \{(-1 \pm 3\sqrt{-3})/2\}$ then $F_t(X, Y)$ is the cube of a linear polynomial. Thus (2) has infinitely many solutions (x, y) for all roots of unity $\mu \in \mathbb{Q}(\sqrt{-3})$ in this case.

Table 3. Sporadic solutions to $F_t(x,y) = 1$ for $\operatorname{Re} t = -1/2$. The solutions to $F_t(x,y) = -1$ are the negatives of the listed values. There are no solutions to $F_t(x,y) = \mu$ for roots of unity μ other than for $\mu \in \{-1, 1\}$ for $\operatorname{Re} t = -1/2$

t	x	y
$(-1 \pm \sqrt{-3})/2$	$\pm 3\sqrt{-3}$	$(1 \pm 3\sqrt{-3})/2$
$(-1 \pm \sqrt{-3})/2$	$(-5 \pm \sqrt{-3})/2$	$-2 \pm \sqrt{-3}$
$(-1 \pm \sqrt{-3})/2$	$(5 \pm \sqrt{-3})/2$	$(-9 \pm 3\sqrt{-3})/2$
$(-1 \pm \sqrt{-3})/2$	$-2 \pm \sqrt{-3}$	$(9 \pm 3\sqrt{-3})/2$
$(-1 \pm \sqrt{-3})/2$	$2 \pm \sqrt{-3}$	$(5 \pm \sqrt{-3})/2$
$(-1 \pm \sqrt{-3})/2$	$(-9 \pm 3\sqrt{-3})/2$	$2 \pm \sqrt{-3}$
$(-1 \pm \sqrt{-3})/2$	$(-1 \pm 3\sqrt{-3})/2$	$\pm 3\sqrt{-3}$
$(-1 \pm \sqrt{-3})/2$	$(1 \pm 3\sqrt{-3})/2$	$(-1 \pm 3\sqrt{-3})/2$
$(-1 \pm \sqrt{-3})/2$	$(9 \pm 3\sqrt{-3})/2$	$(-5 \pm \sqrt{-3})/2$
$(-1 \pm \sqrt{-7})/2$	$\pm \sqrt{-7}$	$(-1 \pm \sqrt{-7})/2$
$(-1 \pm \sqrt{-7})/2$	$(-1 \pm \sqrt{-7})/2$	$(1 \pm \sqrt{-7})/2$
$(-1 \pm \sqrt{-7})/2$	$(1 \pm \sqrt{-7})/2$	$\pm \sqrt{-7}$
$(-1 \pm \sqrt{-19})/2$	$\pm \sqrt{-19}$	$(-3 \pm \sqrt{-19})/2$
$(-1 \pm \sqrt{-19})/2$	$(-3 \pm \sqrt{-19})/2$	$(3 \pm \sqrt{-19})/2$
$(-1 \pm \sqrt{-19})/2$	$(3 \pm \sqrt{-19})/2$	$\pm \sqrt{-19}$
$(-1 \pm \sqrt{-31})/2$	$\pm \sqrt{-31}$	$(-19 \pm \sqrt{-31})/2$
$(-1 \pm \sqrt{-31})/2$	$(-19 \pm \sqrt{-31})/2$	$(19 \pm \sqrt{-31})/2$
$(-1 \pm \sqrt{-31})/2$	$(19 \pm \sqrt{-31})/2$	$\pm \sqrt{-31}$
$(-1 \pm \sqrt{-35})/2$	$\pm 2\sqrt{-35}$	$24 \pm \sqrt{-35}$
$(-1 \pm \sqrt{-35})/2$	$-24 \pm \sqrt{-35}$	$\pm 2\sqrt{-35}$
$(-1 \pm \sqrt{-35})/2$	$24 \pm \sqrt{-35}$	$-24 \pm \sqrt{-35}$

Acknowledgment

The first and the third author were supported by the Austrian Science Foundation FWF, projects S9606 and S9603, respectively, that are part of the Austrian National Research Network "Analytic Combinatorics and Probabilistic Number Theory." Research was partly done during a visit of the first and the third author at the Department of Computer Science of the University of Debrecen in the frame of a joint Austrian-Hungarian project granted by the Austrian Exchange Service ÖAD (No. A-27/2003) and the Hungarian Tét foundation. They thank the department for its hospitality. Other parts were done during a visit of the first author at the Institute of Mathematics of the University of Zagreb in the frame of a joint Austrian-Croatian project granted by the Austrian Exchange Service ÖAD (No. 20/2004 and 23/2006) and the Croatian Ministry of Science, Education and Sports. He thanks the institute for its hospitality. The second author was partially supported by the Hungarian National Foundation for Scientific Research Grant No. T42985. The authors thank Volker Ziegler for pointing out the mistakes in their original paper [6].

References

[1] BILU, YU., HANROT, G. (1996) Solving Thue equations of high degree. J. Number Theory **60**: 373–392

[2] HEUBERGER, C. (2006) All solutions to Thomas' family of Thue equations over imaginary quadratic number fields. J. Symbolic Comput. **41**: 980–998

[3] HEUBERGER, C. (2006) Parametrized Thue equations – A survey. Proceedings of the RIMS Symposium "Analytic Number Theory and Surrounding Areas", Kyoto, Oct. 18–22, 2004. RIMS Kôkyûroku **1511**: 82–91

[4] HEUBERGER, C. (2000) On general families of parametrized Thue equations. In: HALTER-KOCH, F., TICHY, R. F. (eds.) Algebraic Number Theory and Diophantine Analysis. Proc. Int. Conf. held in Graz, Austria, August 30 to September 5, 1998, pp. 215–238. Walter de Gruyter, Berlin New York

[5] HEUBERGER, C. (2006) All solutions to Thomas' family of Thue equations over imaginary quadratic number fields. Online resources. Available at http://www.opt.math.tu-graz.ac.at/~cheub/publications/thuerel-hyper-online.html

[6] HEUBERGER, C., PETHŐ, A., TICHY, R. F. (2002) Thomas' family of Thue equations over imaginary quadratic fields. J. Symbolic Comput. **34**: 437–449

[7] JADRIJEVIĆ, B., ZIEGLER, V. (2006) A system of relative Pellian equations and a related family of relative Thue equations. Int. J. Number Theory **2**: 569–590

[8] MIGNOTTE, M. (1993) Verification of a conjecture of E. Thomas. J. Number Theory **44**: 172–177

[9] THOMAS, E. (1990) Complete solutions to a family of cubic Diophantine equations. J. Number Theory **34**: 235–250

[10] THUE, A. (1909) Über Annäherungswerte algebraischer Zahlen. J. Reine Angew. Math. **135**: 284–305

[11] ZIEGLER, V. (2005) On a family of cubics over imaginary quadratic fields. Period. Math. Hungar. **51**: 109–130

[12] ZIEGLER, V. (2006) On a family of relative quartic Thue inequalities. J. Number Theory **120**: 303–325

Authors' addresses: C. Heuberger, R. F. Tichy, Institut für Mathematik, Technische Universität Graz, Steyrergasse 30, 8010 Graz, Austria. E-Mail: clemens.heuberger@tugraz.at, tichy@tugraz.at; A. Pethő, Department of Computer Science, University of Debrecen, P.O. Box 12, H-4010 Debrecen, Hungary. E-Mail: pethoe@inf.unideb.hu.

Anzeiger Abt. II (2006) 142: 9–11

Anzeiger
Mathematisch-naturwissenschaftliche Klasse Abt. II
Mathematische, Physikalische und Technische Wissenschaften

Electron Density Distribution in the Topside Ionosphere of Mars and Associated Atmospheric Parameters

By

Siegfried J. Bauer

(Vorgelegt in der Sitzung der math.-nat. Klasse am 23. März 2006
durch das w. M. Siegfried J. Bauer)

Radio occultation observations and more recently radar sounding are now providing a large number of electron density profiles of the Martian ionosphere [1–3]. With the demise of the Japanese Nozomi mission, aeronomical parameters for interpreting the Martian ionosphere are thus limited to the early observation on the American Viking mission [4].

That the principal ionospheric layer on Mars corresponds to a photo-chemical Chapman-type F_1 layer is now well established since its peak properties exhibit the expected solar control [5–7]. However the ionosphere of Mars (as that of Venus) deviates from a simple Chapman-layer, since its principal ion (O_2^+) does not correspond to the ionizable constituent (CO_2), but results from an ion-atom interchange reaction with atomic oxygen originating in dissociative processes.

It was shown by the Viking RPA experiment [4] that above the ionospheric peak the lighter ion O^+ is also present in addition to O_2^+. The distribution of O^+ showing a peak near 200 km, seems to be the result of photochemistry below and plasma diffusion above it. Below the O^+ peak, this ion results from the equilibrium of photoionization of O and chemical loss via an ion-molecule reaction involving CO_2, i.e., $O^+ \propto n(O)/n(CO_2)$ exhibiting an increase with altitude according to $\exp(z/H(28))$, i.e., corresponding to an effective mass $m(CO_2) - m(O)$ in the scale height. Above the peak, O^+ is governed by plasma diffusion, although affected by the presence of the major ion O_2^+. Thus, the O^+ distribution on Mars represents a "hidden" F_2 layer [8].

In the topside ionosphere the electron density distribution results from the ion species O_2^+ and O^+ in diffusive equilibrium. In contrast to neutral species, ion species in diffusive equilibrium are not independent of each other because of a polarization electric field set up to prevent charge separation. This field can counteract gravity for a lighter minor ion (m_2) since it depends on the mean ion mass determined by the heavier major ion (m_1) and the sum of the temperature of the electrons T_e and ions T_i that generally exceed the neutral gas temperature [8].

For a binary ion mixture with masses m_1 and m_2 controlled by plasma diffusion, the electron density distribution was first derived by the author four decades ago [9] as

$$N(z) = N_0 \exp\left[-\frac{1}{1+\epsilon}\left(\frac{z}{H_1}\right) - \ln\left(1 + \eta \exp\left(\frac{z}{H_{12}}\right)\right) + \ln(1+\eta)\right],$$
$$(1)$$

where $\epsilon = T_e/T_i$, $H_1 = kT_i/m_1 g$, $H_{12} = kT_i/(m_1 - m_2)g$ and $\eta = N(X_2^+)/N(X_1^+)$ is the ion abundance ratio at the reference level $z = 0$.

In the case of Mars $X_1^+ \equiv O_2^+$ and $X_2^+ \equiv O^+$ representing the appropriate binary ion mixture. For illustrative purposes Fig. 1 shows the topside electron density distribution according to Eq. (1) for different

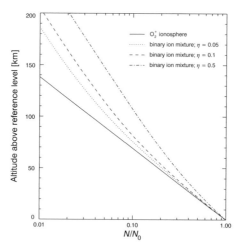

Fig. 1. Topside electron density distribution of a binary ion mixture $(O_2^+ + O^+)$ for various ion abundance ratios η at $z = 0$

ion abundance ratios η at the reference level $z = 0$, with $H_1 = 15\,\text{km}$ and $H_{12} = 30\,\text{km}$, assuming thermal equilibrium ($\epsilon = 1$). It is obvious that fitting observed electron density profiles with the distribution according to Eq. (1) should allow aeronomic parameters to be inferred from topside profiles above an altitude where plasma diffusion is expected to prevail. A reference level $z = 0$ at altitudes $h \geq 180\,\text{km}$ appears to be an appropriate choice. Thus, T_i from scale height, absence of thermal equilibrium ($\epsilon > 1$) and the abundance ratio of neutral constituents O and CO_2 from η may be obtained by fitting observed topside profiles, providing useful constraints for models of the upper atmosphere of Mars [10].

Acknowledgement

The help of M. Rieger in the preparation of Fig. 1 is gratefully acknowledged.

References

[1] HINSON, D. P., SIMPSON, R. A., TWICKEN, J. D., TYLER, G. L., FLASAR, F. M. (1999) Initial results from the radio occultation measurements with Mars Global Surveyor. J. Geophys. Res. **104**: 26997–27012

[2] PÄTZOLD, M., TELLMANN, S., HÄUSLER, B., HINSON, D., SCHAA, R., TYLER, G. L. (2005) A sporadic third layer in the ionosphere of Mars. Science **310**: 837–839

[3] GURNETT, D. A., KIRCHNER, D. L., HUFF, R. L., MORGAN, D. D., PERSOON, A. M., AVERKAMP, T. F., DURU, F., NIELSEN, E., SAFAEINILI, A., PLALUT, J. J., PICARDI, G. (2005) Radar soundings of the ionosphere of Mars. Science **310**: 1929–1933

[4] HANSON, W. B., SANATANI, S., ZUCCARO, D. R. (1977) The Martian ionosphere as observed by the Viking retarding potential analyzers. J. Geophys. Res. **82**: 4357–4363

[5] HANTSCH, M., BAUER, S. J. (1987) Solar control of the Mars ionosphere. Planet. Space Sci. **38**: 3539–3542

[6] BREUS, T. K., KRYMSKII, A. M., CRIDER, D. H., NESS, N. F., HINSON, D., BARASHYAN, K. K. (2004) Effect of solar radiation in the topside atmosphere/ionosphere of Mars: Mars Global Surveyor observation. J. Geophys. Res. **109**: A09310

[7] RISHBETH, H., MENDILLO, M. (2004) Ionospheric layers of Mars and Earth. Planet. Space Sci. **52**: 849–852

[8] BAUER, S. J., LAMMER, H. (2004) Planetary Aeronomy. Springer, Berlin Heidelberg New York

[9] BAUER, S. J. (1962) The electron density distribution above the F_2 peak and associated atmospheric parameters. J. Atmos. Sci. **19**: 235–240

[10] BOUGHER, S. W., ENGEL, S., HINSON, D. P., MURPHY, J. R. (2004) MGS radio science electron density profiles: Interannual variability and implications for the Martian neutral atmosphere. J. Geophys. Res. **109**: E03010

Author's address: Dr. S. J. Bauer, Institut für Physik/IGAM, Karl-Franzens-Universität Graz, 8010 Graz, Austria. E-Mail: siegfried.bauer@uni-graz.at.